W9-DCJ-721

Understanding Optical Fiber Communications

For a complete listing of the *Artech House Optoelectronics Library*,
turn to the back of this book.

MONTGOMERY COLLEGE
ROCKVILLE CAMPUS LIBRARY
ROCKVILLE, MARYLAND

Understanding Optical Fiber Communications

Alan Rogers

Artech House
Boston • London
www.artechhouse.com

266675

MAY 21 2002

Library of Congress Cataloging-in-Publication Data
Rogers, A. J.
    Understanding optical fiber communications / Alan Rogers.
        p.   cm.—(Artech House optoelectronics library)
    Includes bibliographical references and index.
    ISBN 0-89006-478-4 (alk. paper)
    1. Optical communications.   2. Fiber optics.   I. Title.   II. Series.
    TK5103.59 .R64   2001
    621.382'7—dc21
                                                                00-068930
                                                                CIP

British Library Cataloguing in Publication Data
Rogers, A. J. (Alan John), 1936–
    Understanding optical fiber communications. — (Artech House
    optoelectronics library)
    1. Optical communications   2. Fiber optics
    I. Title
    621.3'827
    ISBN 0-89006-478-4

Cover design by Yekaterina Ratner

© 2001 ARTECH HOUSE, INC.
685 Canton Street
Norwood, MA 02062

All rights reserved. Printed and bound in the United States of America. No part of this book may be reproduced or utilized in any form or by any means, electronic or mechanical, including photocopying, recording, or by any information storage and retrieval system, without permission in writing from the publisher.

All terms mentioned in this book that are known to be trademarks or service marks have been appropriately capitalized. Artech House cannot attest to the accuracy of this information. Use of a term in this book should not be regarded as affecting the validity of any trademark or service mark.

International Standard Book Number: 0-89006-478-4
Library of Congress Catalog Card Number: 00-068930

10 9 8 7 6 5 4 3 2

To my wife, Wendy,
and my two sons, Daniel and Gareth

Contents

Preface

Optical-fiber telecommunications technology is steadily changing our lives. Optical fibers, which are being installed under the ground, on overhead lines, and under the oceans, are not immediately visible, but they are providing the means whereby enormous quantities of information, of all kinds and for all types of use, can now be made available to everyone, worldwide. This will change the way we work, play, and generally live our lives.

The technology of optical-fiber telecommunications is developing rapidly. In order that the development should be of maximum benefit to all in society, and should take place in directions controlled by society, it is necessary for the technology to be understood by a large cross-section of people. This includes individuals who have some association with the industry (not necessarily of a directly technical nature) but also those who might later wish to enter the industry (i.e., those at school or in colleges and universities), and those members of the general public who just like to know how things work. This book attempts to address the needs of all these groups.

The book explains, in terms that require no formal scientific or mathematical training, the principles upon which modern optical communications systems are based. It deals in turn with all aspects of the subject: the nature and measure of information, the nature of light and the way in which it passes down optical fibers, the laser sources and receivers of light signals, and the optical amplifiers, modulators, and switches that control the light. This involves a wide range of ideas, which are all brought together in a discussion of system design in Chapter 6. Chapter 7 explains some of the more advanced ideas being pursued at the front edge of the technology for the evolving systems. The final chapter takes a brief, speculative look at what might lie in the future.

There are five appendixes that deal in more formal detail with important topics. These are more mathematical in nature, and are included for the benefit of those readers who have some mathematical background, and who thus might appreciate seeing these ideas expressed in more precise language. The appendixes are not necessary for the overall understanding of the book, and can be ignored by those readers whose interest is of a more general nature.

Devising an expression of the principles of optical-fiber telecommunications in essentially nonmathematical terms has been, for me, a useful and a beneficial exercise. I am much the wiser for it.

In the preparation of this book, I have been helped by several people. My grateful thanks are due to my wife, Wendy, and my two sons, Daniel and Gareth, for helping to prepare the manuscript and for commenting on it in a variety of useful ways. I am grateful also to Professor Brian Culshaw for his technical comments on the text, from which I have benefited considerably. Finally, I wish to thank the staff at Artech House for their tolerance and cheerful assistance during the writing and publishing of this book.

1

What Are Telecommunications?

1.1 The Global Village

Ever since animal life began on our planet, information has been essential to survival. Information is knowledge, and knowledge can be used to sustain life: knowledge about food supplies, the approach of predators, the location of hazards, and so on.

As the evolutionary process advanced, it soon became clear that an effective survival ploy was to get organized into groups: shoals, herds, tribes. In this way, tasks could be shared: a few could watch while others slept. The effective operation of groups required a sharing of knowledge and intelligence, within the group and between groups. It became necessary to transfer information from individual to individual, group to group, place to place. The art of communication thus became important, and the more accomplished was the group in the practice of the art, the more likely it was to survive.

Human beings communicate with each other constantly. As a direct result of the survival advantages of communication, we have become social creatures. One of the primary reasons for our dominance in the animal kingdom is that we are so good at communication. It has enabled us to organize ourselves, beyond groups and tribes, into societies, civilizations, and nations. This book is concerned with one of the most advanced forms of modern communications, one that uses light.

It is often said these days that we live in a global village. What is meant by this is that almost anything of any importance which occurs on one part of the planet can be known, and influence any other part; the reason for this lies in our technology, much of which we take very much for granted in the present day and age. This is nicely illustrated by the (reportedly) true story of the elderly lady who was invited by her son to fly across the United States to watch her grandson play in a major-league baseball game. "Oh, no," she said, "I'm not traveling on one of those airplanes; I'll stay at home and watch him on television as the good Lord intended."

It is sobering to be reminded that it is only 120 years since Alexander Graham Bell invented the telephone (Figure 1.1), and only 150 years since

Figure 1.1   Alexander Graham Bell. (Photograph from the Archives of the Institution of Electrical Engineers.)

Samuel Morse invented the telegraph. Prior to 1844 there were no electronic means of communicating at all; any "out of sight" communication relied on the physical transport of (usually) handwritten messages, often via horseback or wagon wheel. Such communication can be most complete, but it is unreliable and, of course, very slow. There have been great technological advances in the last few generations of human development, in almost all areas of endeavor.

Telecommunication literally means "communication from afar," and from very early times mankind recognized its advantages in both war and peace. In this, speed of communication was understood to be probably the most important feature, for in the vast majority of cases, some action had to be taken before it was too late: defenses had to be mobilized against attack; cities had to be evacuated before the flood, and so on.

In many early methods of telecommunication, light was used: the wave of the hand, the heliograph (signaling with mirrors that reflected the sun's light), burning beacons spaced at regular intervals between communication stations (Figure 1.2), the Aldis lamp from ship to ship. Sound was also used—with the "bush telegraph" in the jungles of Africa, for example—but it was not until electricity was brought into the arena that things really began to take off.

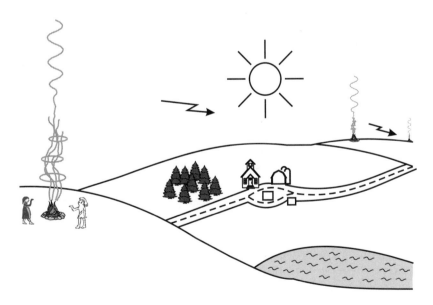

Figure 1.2 Communication via hilltop beacon.

This takeoff point marked a fundamental change not only in the nature of the technology but also in its effect on the way we live our lives. Prior to 1850 communications advances within and between societies, slow and cumbersome as they were, helped the societies to become more efficient, to improve standards of living, and to develop knowledge bases, but the societies themselves remained broadly the same. With the advent of electronic means of communication, the very structure and fabric of society began to be altered and conditioned by that technology. It is for this reason that many controlling autocrats (dictators and cliques) have been wary, even openly hostile, to the free communication between individuals which telecommunications technology provides: Adolf Hitler discouraged telephone development in Germany during the Nazi era, as did Joseph Stalin in the Soviet Union. Totalitarianism requires control over people's minds which freedom of information makes impossible.

On the other hand, the glory of noninteractive television is a powerful tool in the hands of a ruling clique, for it can be used as a means for brainwashing and conditioning, via the broadcast of carefully selected propaganda. George Orwell saw this very clearly, and gave powerful expression to it, in his novel *1984*, where television was used as just such a tool; fortunately, Orwell's predictive powers were somewhat wide of the mark, and 1984 was a more congenial year than he envisaged. On the more positive side, mass telecommunication has improved enormously the economic effectiveness and standards of living in those modern societies that can presently afford it. Business efficiency, educational methods, entertainment, and human understanding have all improved by giant leaps with each major new development in technology. The most recent of these, the global Internet, promises to alter mind-sets yet again, rendering professions, institutions, and governments incapable of withholding information from the general population, and hence altering the very structure of our political processes and of democracy itself. In addition, the world's knowledge base becomes much more freely and beneficially available to all.

The impact, then, of telecommunications development for all of us, and for all our children, is considerable. As with all science and technology, the way to control it and to ensure that it is used for the benefit, rather than to the detriment, of humankind is to understand it. Optical-fiber communication is set to become the dominant telecommunications technology of the twenty-first century. In order to fully appreciate its potential, and how best to use it, we therefore need to understand it. The purpose of this book is to spread the elements of that understanding as widely as possible beyond the specialists in the subject.

We begin by looking at the concepts that are valuable to telecommunications as a whole, in order to obtain a clear picture of what the word means, what methods and processes it employs, and how it has developed. This is valuable knowledge in itself, but it also allows us to appreciate why we have been led to optics, and the place optics has in the broader scheme of things.

1.2 Elements of telecommunications

In order to convey information from one place to another, we need a communications system.

The basic communications system consists of three components: a transmitter, a communications channel, and a receiver (Figure 1.3). Almost all books on elementary telecommunications begin in this way, and this book will be no exception. All communications systems conform to this basic structure.

When one person talks in normal conversation with another, the speaker is the transmitter, the air is the channel, and the listener is the receiver. When one person waves his hand to another, the hand is the transmitter (of reflected light), the space between the two people is the channel, and the eye of the other person is the receiver. When a fire sets off a fire alarm in a building, the smoke sensor is the transmitter, a pair of copper wires is the channel, and the audible (or other) alarm is the receiver: this then acts as the transmitter for a second system, with air again as the channel and all persons within earshot as the receivers.

An appreciation of this simple arrangement immediately points to a number of general questions that must be answered by the communications system designer:

- In what form is the information that is to be conveyed?

- How can this information be used by the transmitter?

Figure 1.3 Basic telecommunications systems.

- How is the information fed by the transmitter to the communications channel?

- What effect does the channel have on the information?

- How does the receiver extract the information from the channel?

- In what form should the receiver present the information to the outside world?

- How does the received information differ from the original information? What causes the differences? To what extent can the two be *allowed* to differ?

The full answers to these questions would take us deeply into telecommunications theory and practice. However, a good appreciation of most of the answers can be obtained via a familiarization with some quite simple ideas.

1.3 Information

Several times now we have met the word *information* in the loose context of "intelligence" or "knowledge." It is information that is fed to the transmitter and information that is retrieved by the receiver, for example. In science and technology it is impossible to treat any ideas seriously unless they deal with things that can be *measured.* How can we decide, for example, how well a telecommunications system is performing? Its function is to convey information between two separated points, and we can only know how well it is doing that if we have some way of measuring the information in, at one point, and the information out, at the other. None of the questions posed in the last section can be answered in a useful (to the system designer) way unless we have a means of measuring information. So how do we measure information?

Information theory seeks to answer this question, among others, and is a highly mathematical subject. However, the essence of the ideas involved is simply stated, and in a way that is sufficient for our present purposes.

It is only relatively recently (circa 1948) that information theory was put on a firm base by the American mathematician Claude Shannon. In seeking a measure of information, Shannon drew on the experience of playing the well-known parlor game "twenty questions". Here, as many readers will know, one person thinks of a particular item, such as a window or a violin, and the team players try to discover what the item is by asking that person a maximum of twenty yes or no questions. The challenge for the questioners is to plan their questions so as to maximize the information provided by each yes or no answer. For example, in the case of a window:

1. Is it a vegetable?	*No.*
2. Is it a mineral?	*Yes.*
3. Is it a useful object?	*Yes.*
4. Is it used in sport?	*No.*
5. Is it used in the house?	*Yes.*

And so on. Shannon realized that any piece of information can be broken down into a series of such yes or no answers, and that the best measure of the information content is the minimum number of questions needed to represent the information. Since each of the questions has only one of two answers, the two answers can be represented by one of two digits: 0 for no and 1 for yes, say. These are called binary digits, or bits. Hence the information content in a message can be expressed as the number of bits corresponding to the number of binary (i.e., there are two possible answers) questions that must be asked in order to convey the information. Suppose, as an example, we wish to measure the number of bits present in the information represented by the sentence *the cat sat on the mat.* This sentence consists of a series of letters, spaces, and punctuation marks. Let us call all of the possible positions for these elements "slots." Each slot might thus contain one of 26 letters, or one space, or one of (say) 5 punctuation marks (comma, period, semicolon, etc.), making 32 possibilities in all. If these various possible symbols are each assigned numbers from 1 to 32 (Figure 1.4), then the first letter of the sentence, *t*, turns out to be number 20 in the alphabet, and we might proceed to ask (and answer) our binary questions as follows:

1. Does the first symbol lie between numbers 1 and 16 (inclusive)?	*No* (We now know it lies between 17 and 32)	Bit representation: (0)
2. Does it lie between 17 and 24?	*Yes*	(1)
3. Does it lie between 17 and 20?	*Yes*	(1)
4. Does it lie between 17 and 18?	*No*	(0)
5. Is it 20?	*Yes*	(1)

Therefore we have a binary code for *t*: 01101.

This process is illustrated in Figure 1.4. We have identified the first letter of the sentence by using just five binary questions, and thus the first letter effectively contains 5 bits of information (01101). Clearly, each subsequent symbol can be established in exactly the same way, by asking just five questions, so that each slot contains 5 bits of information. Since there is a total of 23 slots in the sentence, its total information content is $5 \times 23 = 115$ bits.

A little thought will reveal that what we have done in this case is to halve the range of possibilities with each question asked (see Figure 1.4). In order to identify one symbol from 32 possibilities we have had to do this five times, because

$$\frac{1}{2} \times \frac{1}{2} \times \frac{1}{2} \times \frac{1}{2} \times \frac{1}{2} = \frac{1}{32}$$

In other words, there are 5 bits because:

$$2 \times 2 \times 2 \times 2 \times 2 = 2^5 = 32$$

(Note that when 2 is multiplied by itself five times, we represent it as 2 to the fifth power; i.e., 2^5. Any number can be represented similarly, e.g., $1,000 = 10 \times 10 \times 10 = 10^3$).

Clearly, the number of bits in this case is the power to which the number 2 (in our binary system) must be raised in order to correspond to the number of slots.

If each of the possibilities is not equally probable (as indeed is the case for *the cat sat on the mat*), since some letters of the alphabet are used more frequently than others, or some are more likely than others to follow a given letter, then these arguments must be refined somewhat. We might argue, for example, that, having established the first three words as *the cat sat*, there is a good chance that the next word is *on,* because this word very often does follow the word *sat.* Using prior knowledge about message content in this way, it is possible to use fewer bits than would otherwise be the case, a process known as compression. This is a very important topic in modern telecommunications systems, because the degree of compression, and therefore the communications advantage, can be very great. The basic ideas, however, remain the same.

With this measure of information, we now have the means by which we can determine how well a communications system is working, and how

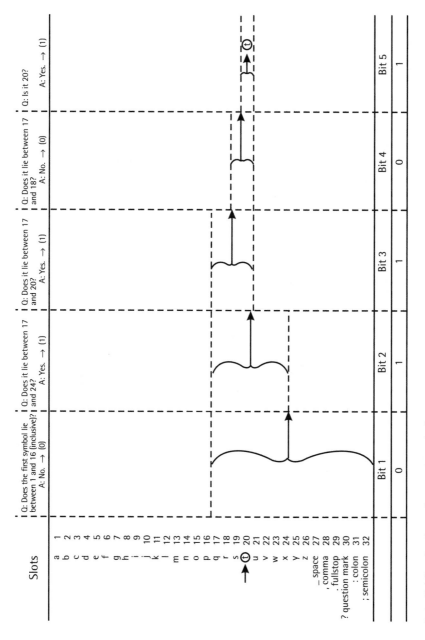

Figure 1.4 5-bit selection of correct symbol.

one system can be compared with another, by measuring how much information is communicated between separated points in a given time. Clearly, the more information that is communicated in a given time over a given distance, the more powerful is the system.

We must now look in more detail into how, exactly, communications systems work.

1.4 Signals

First we will look more deeply into the mechanisms by which information is communicated. In all cases, communication is effected by varying the size of a convenient physical quantity in a way that is uniquely related to the information. Take the example of normal speech. When one person talks to another, the information to be transmitted originates in the speaker's brain, is conveyed to the vocal chords, which encode the information by causing corresponding changes in pressure on the air within the mouth, leading to pressure (sound) waves, which propagate through the air to the ears of the listener. The ears then decode the signal (via the action of the ear's "hammer" on its "anvil") for interpretation by the listener's brain. Human evolution has arranged that this is a very complete and efficient communication system, but only for short distances. In this case the physical quantity being varied is the pressure of the air. If the pressure of the air were measured in the vicinity of the listener's ear, it might vary with time as shown in Figure 1.5b. This variation we would call a signal. Clearly, it contains information.

Similarly, if we were to measure the electric current flowing into a telephone handset during a telephone conversation, this current signal would be of the same form as the air pressure signal in Figure 1.5b. In fact, all telecommunications systems in use today, and into the foreseeable future, make use of electrical signals, so we must understand more about these.

Electrons are very small elementary particles that carry a small negative electric charge. They comprise one of the basic building blocks of atoms. Electric charges of the same sign repel each other, and this force of repulsion can lead to movement of the electrons, which is called an electrical current. The repulsive force that causes the electrons to move is generated by a voltage, which is a measure of the total effect of this force acting between two points, a kind of electrical pressure. So we have an electrical pressure (a voltage), causing a flow of electrons (a current). Currents flow

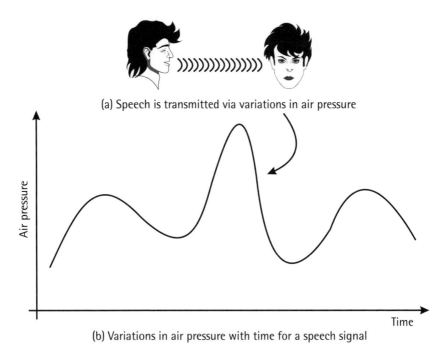

(a) Speech is transmitted via variations in air pressure

(b) Variations in air pressure with time for a speech signal

Figure 1.5 Speech signals.

very easily in good conductors, metals such as copper or silver. Copper is much cheaper than silver and so is almost universally used to carry currents over long distances. We can communicate easily using current signals because we can supply voltages between the ends of two long parallel copper wires, and the resulting currents will flow in the wires for very great distances. They can then be converted back into voltages at the far end of the pair, by measuring the force of the flow. Usually this is done by passing the current through an electrical resistance that effectively senses the strength of the flow and then generates a corresponding voltage as a result. These voltages will be similar in shape to the original driving voltages at the front end. If, therefore, we can convert speech pressure waves into a voltage, we can send speech information down the pair of wires: this is what a microphone does. If we can convert the voltage at the far end of the wires back into sound, we can then listen to the original sound at the far end of the wires: this is what an audio earpiece or a loudspeaker does. What has just been described is a telephone system; this is precisely how a telephone works (see Figure 1.6).

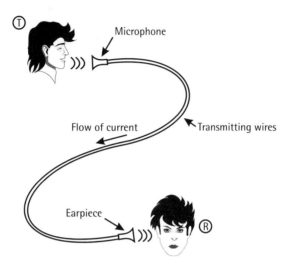

Figure 1.6 Basic telephone system.

It is convenient to convert information into electrical signals for two main reasons: first, electrical currents will flow with little change over very long lengths in a good conductor, so that the information will be accurately preserved over these distances; second, the science and technology of electronics has advanced to the point where we can control electrons with great ease and precision. This is enormously useful in the design and construction of complex telecommunications systems.

Thus we find ourselves dealing very often with voltage signals, which is to say that the physical quantity that represents the information we wish to convey is expressed as a variation of voltage (electrical pressure) with time (Figure 1.7a). The important question we must now ask is: How can we work out the *quantity* of information in such a signal; that is, how many bits does it contain?

To answer this question it is necessary, of course, to revert back to the idea of our binary (yes or no) questions. But how do we ask such questions of this continuous variation of voltage? We need to employ a rather devious stratagem in order to answer this.

Suppose that we look at the waveform at particular times and that these times are regularly spaced along the signal (Figure 1.7a). At each of these times we measure the size of the signal and record the values. In more technical language we would say that the signal "amplitude" had been "sampled" at regular intervals.

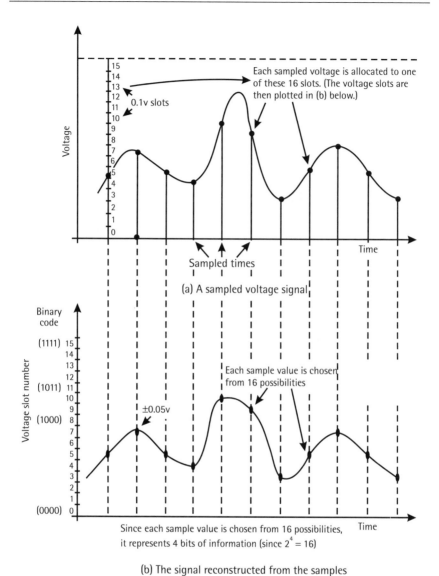

Each sampled voltage is allocated to one of these 16 slots. (The voltage slots are then plotted in (b) below.)

0.1v slots

Voltage

Sampled times

(a) A sampled voltage signal

Binary code

Voltage slot number

(1111)

(1011)

(1000)

±0.05v

Each sample value is chosen from 16 possibilities

(0000)

Time

Since each sample value is chosen from 16 possibilities, it represents 4 bits of information (since $2^4 = 16$)

(b) The signal reconstructed from the samples

Figure 1.7 Signal sampling and reconstruction.

If we now plot those sampled values, including some measurement error, and join them together reasonably smoothly, the result bears a fair resemblance to the original signal. This is illustrated in Figure 1.7b. It is clear that the resemblance will be closer the more often the signal waveform is sampled and the more accurately the amplitudes at the sampling points are measured.

Now, there will always be an error in any measurement process. It is impossible to know the value of any physical quantity with absolute precision. This is because there is always some unwanted signal variation or "noise" present. (We will look at this in more detail later.) So suppose that, in this case, the noise prevents us from measuring the amplitudes to better than 0.1 volt (v) (this means that the sampled amplitude would be stated in the form $1.2 \pm 0.05v$). Suppose also that the maximum amplitude the signal can reach is 1.6v. It is then convenient to break this 1.6v down into a set of voltage intervals, each 0.1v wide, making a total of 16 of them (Figure 1.7a). When we measure the amplitude now, essentially what we are doing is deciding within which of these intervals the signal level lies. Each of the 16 intervals is given a number, from 0 to 15, say. Hence we now have a situation very similar to that of *the cat sat on the mat*. The first binary question is: Does the signal amplitude lie within an interval whose number is in the range 0 to 7? If no, does it lie in the range 8 to 11? And so on. Since there are now 16 possibilities, it is clear that 4 such questions will establish the interval in which the level lies, and thus will establish the voltage level to accuracy $\pm 0.05v$. Hence the information in that signal level is equivalent to 4 bits. Again we note that the number of bits is the power to which 2 (it is still a binary system) must be raised in order to represent all possibilities, that is:

$$2^4 = 16$$

Each sample can be represented by 4 bits. Another way of looking at this is to understand that each of the 16 numbers from 0 to 15 can now be represented by a binary code. This code simply states which of the four powers of 2 (i.e., 2^3, 2^2, 2^1, 2^0) are to be added to equal any given one of the numbers. For example, for the numbers 0, 8, 11, and 15 we have:

Power of 2:	2^3	2^2	2^1	2^0
	(8)	(4)	(2)	(1)
Number		Binary Code		
0	0	0	0	0
8	1	0	0	0
11	1	0	1	1
15	1	1	1	1

These are shown on the vertical axis in Figure 1.7b.

If there are N samples in the total signal length, each sample having an information content of 4 bits, then the total information content clearly will be 4N. As stated earlier, the accuracy with which the signal is reproduced by this sampling process depends upon the accuracy of measurement of the sampled amplitude and on the rate at which it is sampled, that is, the sampling interval. Of course, we also have to know how accurately the signal *needs* to be reproduced before we can determine these things, and this will depend upon what it is to be used for.

The sampling is done at a certain rate. How do we know what rate this should be? How does this rate affect the representation of the signal? To answer these questions we must now consider the important matter of signal bandwidth.

1.5 Bandwidth

The word *bandwidth* occurs frequently in association with telecommunications systems. It is sometimes used as a synonym for *capacity,* the maximum amount of communications traffic which can be handled by the system. We are now, as a result of the last few sections, able to be more precise, and define the capacity in terms of the number of bits of information that can be handled by the system in, say, one second; and there certainly is a connection between this quantity and the system bandwidth; this will now be explored.

Consider the wave shown in Figure 1.8a. This might represent the variation with time of any physical quantity such as voltage, air pressure, water level, et cetera. It is described completely by an amplitude (the height of a peak), a frequency (the number of complete "cycles," such as the cycle that lies between A and B in Figure 1.8a, in one second) and a phase (the timing of a peak with respect to some reference time, depicted in the figure as $t = 0$). If this wave continues for all time it contains no information, for it is entirely predictable at a receiver, each of its describing features being constant in time. It is a "pure" wave, and tells us nothing other than that it is there.

Suppose however that we were able to switch it on and off. In this case we could transmit information with its aid: for example, in Morse code, with the on state representing a dash and the off state a dot, as shown in Figure 1.8b (or the on state representing a 1 and the off state a 0, in a binary system).

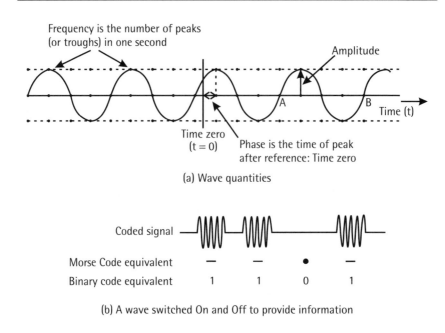

Frequency is the number of peaks
(or troughs) in one second

Amplitude

A

B

Time (t)

Time zero
(t = 0)

Phase is the time of peak
after reference: Time zero

(a) Wave quantities

Coded signal

Morse Code equivalent   —   —   •   —

Binary code equivalent   1   1   0   1

(b) A wave switched On and Off to provide information

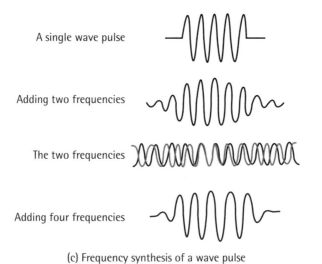

A single wave pulse

Adding two frequencies

The two frequencies

Adding four frequencies

(c) Frequency synthesis of a wave pulse

Figure 1.8 Waves, wave pulses, and wave synthesis.

Let us look now at what this does to the wave. Take the waveform for a
single dash, as shown in Figure 1.8c. This is not now a wave that continues

for all time; it is what is known as a wave "pulse" or, sometimes, a "burst." This pulse can, in fact, be formed from a number of pure waves, as illustrated in Figure 1.8c. As more and more waves with frequencies close to the frequency of the original wave are added, it is possible to arrange for the resultant waveform to approximate more and more closely to the original pulse of waves. (Clearly, they are all "in phase" at the center of the pulse, but not anywhere else). By forcing the signal to provide information, in the form of a dash, we have forced it to contain a spread of frequencies, rather than just the one frequency of the single, pure wave. And the more information we force this wave to carry, the greater the range of frequencies that is required, for our knowledge of the start and stop times of the pulse is improved as more frequencies are added to reproduce it (see Figure 1.8c). Now, the spread of frequencies present in a signal is called its bandwidth. Hence we can see immediately the connection between information content and bandwidth. The process we have just been considering, that of building up a waveform by adding together a number of pure waves of different frequencies, is known as Fourier synthesis (after the famous French mathematician who developed it), and it is a powerful tool in telecommunications theory. Some of the mathematical details of the process are included in Appendix A, for readers who might be interested.

Let us see now how this applies to a more general waveform, such as the speech waveform first shown in Figure 1.5, and shown again in Figure 1.9a. Because this is a continuous waveform and follows accurately the speech pressures produced by the vocal chords, it is called an "analog" waveform, since the variation is analogous to the form of the original information. This is to be compared with information in the form of yes or no answers to questions that, since it requires only two digits, 0 and 1, to represent it (like Morse code for example), is said to be in "digital" form. The words *analog* and *digital* are frequently heard in relation to all kinds of telecommunications matters.

Returning now to the analog speech waveform in Figure 1.9a, the question is: How do we measure its bandwidth? The telecommunications engineer would again use Fourier synthesis. There are mathematical techniques that will provide for him the full frequency range of waves which need to be added to produce the signal, together with their amplitudes and phases. In other words, these techniques provide the full frequency "spectrum". But presently we are only interested in the bandwidth—that is, in the range of frequencies present—and we can get a good feel for this, using qualitative arguments that have no need of the full mathematics.

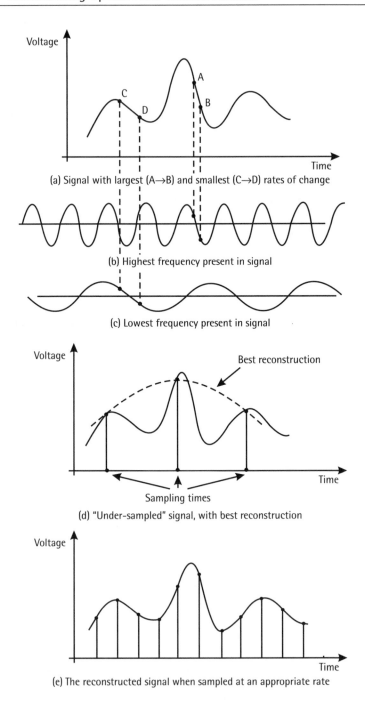

(a) Signal with largest (A→B) and smallest (C→D) rates of change

(b) Highest frequency present in signal

(c) Lowest frequency present in signal

(d) "Under-sampled" signal, with best reconstruction

(e) The reconstructed signal when sampled at an appropriate rate

Figure 1.9 Signal bandwidth and sampling.

Let us look at one small portion of the waveform in Figure 1.9a, that shown between points A and B. This is, clearly, a part of the variation where the rate at which the amplitude is changing is at a maximum. If a detector, such as the human ear, is to respond to this rate of change, it must have a detector mechanism that can change as quickly as this. If it does not have this facility, some information present in the signal will be lost to the ear. Now, if the detector can respond to the rate of change shown between A and B, it also will be able to respond to the pure wave shown in Figure 1.9b, for the maximum rate of change in that wave is the same as that from A to B. Hence there is a very real sense in which the change from A to B effectively implies that a wave with the frequency shown in Figure 1.9b is present in the signal. The Fourier synthesis would show that it was, for there is no way that the waveform could be built up from the sum of other waves unless at least one of the waves was allowing the amplitude to change as fast as the change from A to B.

Look now at the portion of the signal between C and D in Figure 1.9a. This represents the region where the rate of change is at its slowest; that is, a minimum rate. Hence the ear need not respond any more slowly than this in order to receive all the contained information, and thus it need not respond to wave frequencies below the value shown in Figure 1.9c. Hence this wave frequency is the smallest necessary to reproduce the waveform in the Fourier sum. The range of frequencies from the smallest to the largest is the required bandwidth of this signal.

Hence, to summarize, the upper and lower limits of the bandwidth of a signal are telling us about the maximum and minimum rates at which the signal changes. Quite often, the minimum rate will be very close to zero, so that the bandwidth in these cases effectively will be just equal to the highest frequency present.

For the record, and to fix ideas a little, the bandwidth of a typical speech waveform has frequencies in the range from about 300 to 3,000 cycles per second. The human ear can respond up to about 20,000 cycles per second; this enables orchestral music to be well appreciated.

One cycle per second is a unit given the name Hertz, in memory of the German physicist who first demonstrated the existence of radio waves (in 1888). The unit is usually abbreviated Hz, and multiples of 1,000 are used with the well-known prefixes: 1,000 Hz is 1 kilohertz (kHz), 1,000,000 Hz is 1 megahertz (1MHz), and so on. Hence the above frequencies can be written as 300 Hz, 3 kHz, and 20 kHz, respectively.

As people get older, their ears cannot respond as quickly to the incoming waves, so that they cannot hear the high frequencies as well as they could when younger. Their reception bandwidth has become smaller.

Finally, we are now in a position to answer the question posed at the end of Section 1.4; that is, how frequently does an analog waveform need to be sampled in order to reproduce it accurately? Taking the case of a signal that has a bandwidth of zero cycles per second (Hz) to B cycles per second (Hz), it is clear from the above discussion that we will not be able to reproduce the fastest changes unless the sampling is done at least as often as the highest frequency present. If this were not the case we might easily miss these fastest changes as shown in Figure 1.9d. In fact, there is a famous theorem, known as the Nyquist sampling theorem, which proves that the signal must be sampled at a rate equal to at least twice the maximum frequency present; in this case, at 2B samples per second. For the more mathematically minded reader, this is proved in Appendix B. (The reason for the factor of 2 is that it is necessary to know not only the amplitude of the highest frequency present but also its phase, so that two pieces of information are required per cycle of this highest frequency.)

The original signal shown in Figure 1.9a is shown again in Figure 1.9e being sampled at a rate equal to twice the highest frequency it contains (this frequency is that shown in Figure 1.9b). Clearly, the signal can now be reconstructed quite accurately from these samples.

We can now easily calculate the amount of information present in an analog signal. If the accuracy of measurement for each sample allows it to be specified with N bits (e.g., N = 4 for 16 possible levels) and the signal bandwidth is B (e.g., 3 kHz for a speech signal), then the amount of information delivered per second is 2B (Nyquist sample rate) times N (bits per sample), and thus equal to 2BN bits per second. For the speech signal this would be $2 \times 3,000 \times 4 = 24,000$ bits per second, or 24 kb.s^{-1} in conventional notation. In other words, the speaker is providing information at a rate of 24 kb.s^{-1} to the listener.

1.6 Analog and digital systems

In the previous section we learned how to measure the amount of information present in an analog signal. The analog signal was sampled at a rate equal to twice its bandwidth (range of frequencies present) and each sample amplitude was represented by a binary code, to a precision that depended upon the accuracy with which the signal could, or needed to be, specified. We saw also that if N was the number of bits needed to represent the accuracy of each sample (i.e., the accuracy needed was one part in 2^N), then the rate at which the signal was delivering information was 2BN bits per second. Clearly, each bit is either 1 or 0 (yes or no) and hence can be represented by the presence or absence of a pulse. Thus each sample can be represented by a stream of pulses. For example, if the sample amplitude lies within interval 13 (see Figure 1.7), then the pulse stream will be

$$
\begin{array}{cccc}
1 & 1 & 0 & 1
\end{array}
$$

$$
\text{Corresponding to} \quad 2^3 \quad 2^2 \quad (2^1) \quad 2^0
$$

$$
\text{That is,} \quad 8 + 4 + 0 + 1 = 13
$$

If the successive samples, being taken at rate 2B per second, are then all put together as a continuous sequence, we will have a continuous stream of pulses at rate 2BN per second; that is, for the speech sample given as an example in the preceding section, a series of pulses at a rate of 24 kb.s^{-1} (see Figure 1.10). This pulse stream contains all the information present in the analog signal—indeed, we know that it is a direct measure of that information. It follows that, by transmitting the pulse stream rather than the analog signal, we have an alternative way of transmitting that same information.

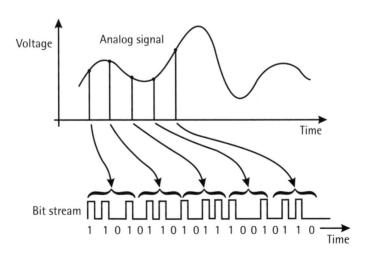

Figure 1.10 A digital bit stream representing an analog signal.

The analog signal is then said to have been converted into digital form: we have digitized the analog signal.

What are the advantages of transmitting the information in this digital form? They are, in fact, considerable, for now all that the receiver in the digital communication system has to do is to detect simple pulses, all of which are the same shape and height. It has only to recognize whether such a pulse is present or absent in any given prescribed time interval. This is a much less demanding task than for it to have to determine, as frequently as it can, which one of many levels the analog signal has.

So we have gained an important advantage by digitizing the analog signal and transmitting the digital pulse stream.

What's the catch? (Nothing comes free of charge!) To answer this question we need to look quite closely into some other related matters concerning the communications channel itself.

All information, analog or digital, has to be transmitted over a channel of some kind to a receiver, so it is necessary to understand exactly how this can be done. We will begin by looking at the way in which the transmitter encodes the information for launching into the channel.

1.7 The transmitter

If we wish to transmit a telephone signal (i.e., a speech waveform) between two points T (transmitter) and R (receiver), say, then our communications system takes a quite simple form, as was depicted in Figure 1.6. A microphone converts the air pressure waves, from a speaker's mouth, into an electric current at T, the electric current passes down a pair of copper wires to an earpiece, at R, which converts the electric current back into sound waves for the listener's ear.

Simple constructions for the microphone and the earpiece are also shown in Figure 1.11. The air pressure variations cause a diaphragm in the microphone to vibrate in sympathy with them; this either compresses or rarefies the density of a block of carbon granules in a container, which varies the electrical resistance (high density, low resistance; low density, high resistance) of a circuit, thus causing variations of current to flow in response to a fixed, applied voltage; again, these variations in current will follow the variations in pressure of the sound wave. The electric current (flow of electrons) passes along the copper wires to the receiver at R where it passes into a coil of wire wrapped around a magnet in the earpiece. This arrangement is known as an electromagnet, and it produces a magnetic

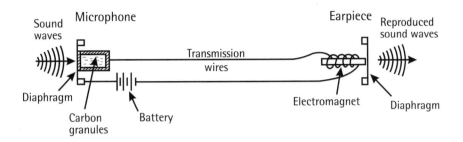

Figure 1.11 Components for a telephone connection.

field, at the magnet's pole-piece, which varies in sympathy with the current flowing through it. This magnetic field exerts a force on an iron diaphragm close to the pole-piece so that the diaphragm moves back and forth in sympathy with the current. The movement of the diaphragm produces a varying pressure on the surrounding air and thus regenerates the original sound wave and passes it into the listener's ear.

A direct connection by means of copper wires in this way is quite convenient for connecting a small number of people to one another, but clearly becomes very quickly impractical when large numbers of people are involved, because it would require a network of copper wires crisscrossing all over the countryside. This raises the question of how one configures a "network," and there are many solutions depending on the particular requirements, but all solutions involve the transmission of many signals down one channel (e.g., a copper wire) at the same time. For example, the "star" network shown in Figure 1.12a shows a network with a switching station, at its center, which is capable of connecting any two of the points it serves (nodes); and Figure 1.12b shows a "star-of-stars" network, which requires connection of these stars to a central station via channels that must carry many signals simultaneously. These latter connections are known as "trunk" lines and the peripheral stars comprise the "local network." Networking is very important and is dealt with as a separate subject in relation to optical systems in Chapter 6. But the first question that arises here is: How can we pass more than one speech channel down one pair of wires at the same time, as is required in a trunk telephone system, in such a way as to enable them to be separated at the receiver end, in other words so that they do not interfere with each other? The answer to this question is fundamental to telecommunications technology. Without an answer we could never have national, international, or global telecommunications.

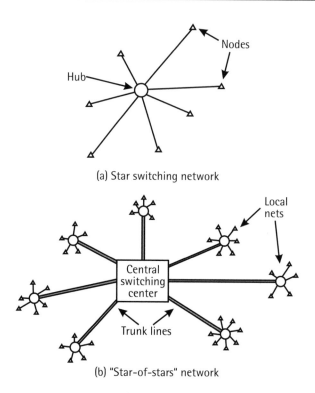

(a) Star switching network

(b) "Star-of-stars" network

Figure 1.12 Switching networks.

The answer is not to transmit the signal as it is but to impress it upon a wave of high frequency; in fact this must be a frequency higher than the bandwidth of the signal (more details of this will be given later). This high-frequency wave is known as the carrier frequency (since it "carries" the information signal) and the process of impressing the signal on to the carrier is known as modulation (since it "modulates" one of the carrier wave's describing features).

The carrier is a pure, simple wave of the type we saw in Section 1.5 (and Figure 1.8a). It contains no information in itself. It is described, as we know, by an amplitude, a frequency, and a phase. To impress the signal information upon it, it is necessary to vary one of these defining quantities in sympathy with the information signal: thus we have amplitude modulation (AM), frequency modulation (FM), or phase modulation (PM), according to which quantity we choose to vary. All types of modulation have their advantages and disadvantages. An example of a carrier wave

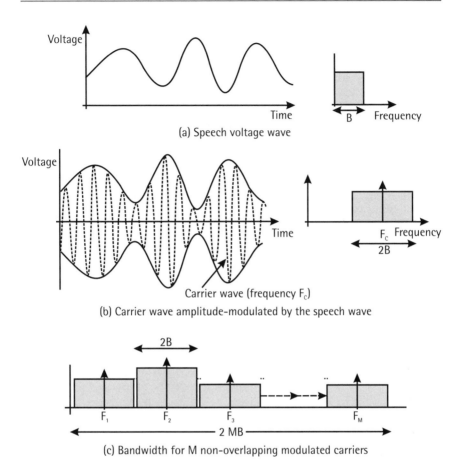

(a) Speech voltage wave

(b) Carrier wave amplitude-modulated by the speech wave

(c) Bandwidth for M non-overlapping modulated carriers

Figure 1.13 Amplitude-modulation bandwidth.

amplitude-modulated by a speech waveform is shown in Figure 1.13b. We know that the carrier, being a single, pure frequency, has no bandwidth: it is at once the highest and the lowest frequency present, and hence the frequency range is (theoretically) zero. However, we also know from Section 1.5 that as soon as it stops being a pure, simple wave, other frequencies are introduced. Hence, as soon as signal information is impressed upon it, as when the pure wave was switched on and off, these other frequencies are introduced. And it will come as no surprise to learn that the range of new frequencies introduced by the modulation process is directly related to the range of frequencies present in the information signal; that is, to the signal bandwidth. The exact relationship between the two depends on the type of modulation we are dealing with (AM, FM, or PM) and a full consideration of this would take us too deeply into information theory for the scope of this book.

However, it is clear that the modulated carrier will have a bandwidth, centered on the frequency of the carrier and extending either side of it. For amplitude modulation (AM) the modulated carrier bandwidth is twice as large as the signal bandwidth (Figure1.13b); if the carrier frequency is F and the signal bandwidth is B, then the bandwidth extends from (F − B) Hz to (F + B) Hz, comprising a bandwidth of 2B Hz (for a mathematical justification of this, see Appendix A). This means of course that the carrier frequency, F, must be larger than B, for F − B cannot be allowed to go negative. (What, indeed, could a negative frequency mean?)

For frequency modulation (FM) and phase modulation (PM), the modulated carrier bandwidths are more than twice as large as for AM, and depend in a more complex way on exactly how the modulation process is carried out.

But how does this provide the answer to our problem of transmitting many signals over the same channel? The answer lies in the use of different carrier frequencies. Several carrier frequencies can be passing down the channel (e.g., a copper wire) at the same time, each carrying its own information signal. These differing frequencies can be recognized separately (i.e., "filtered") at the receiver and then passed into different circuits, a process known as "frequency selection." Each individual carrier then has its own modulation signal removed electronically, a process known as demodulation (since it is the reverse of modulation), thus separately reclaiming all of the original information signals, which are then sent, for the last parts of their journeys, over dedicated copper wires to receivers at their intended destinations.

As already noted, the individual copper wires for these final journeys comprise a local network; the common wires that carry all the modulated carriers simultaneously comprise the trunk network. This trunk network is all-important. It comprises about 99% of all the bandwidth × distance linkage involved in global telecommunications. Clearly, the more carrier signals it can carry, and the greater the distance over which it can carry them, the greater will be the capacity of the network. It is for this reason that the development of telecommunications technology has been, and still is, all about developing long-distance channels capable of carrying more and more carrier frequencies; that is, having more and more bandwidth over greater and greater distances.

In the case of amplitude modulation, we saw that the total bandwidth required to transmit the modulated carrier was 2B Hz. Hence for M carriers, the bandwidth required is M times as large, or 2MB Hz (Figure 1.13c), if they are to remain separate and thus not interfere with each other. Clearly, there must be a limit to the number of modulated carriers a given channel can take. What are the factors that determine this? To answer this question it is necessary to learn more about the actual transmission channels. But before doing this we will take a look at telecommunications receivers, for it is ultimately the receiver in the system that determines whether or not the channel has delivered a satisfactory signal. Hence, before studying the channels themselves, we need to understand something about how the receiver operates.

1.8 The receiver

There are various types of telecommunications receivers, but the most important distinction to be made for our present purposes is between receivers of analog signals and receivers of digital signals. We will deal, in turn, with the essentials of these two types of receiver.

1.8.1 The analog receiver

The function of the analog receiver is to accept the incoming signal from the communications channel and to present the information contained in the signal in a convenient and useful form. The general structure of an analog receiver is shown in Figure 1.14.

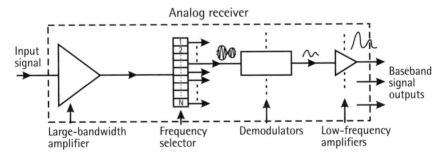

Figure 1.14 Basic analog receiver.

We have already looked at the way in which one type of analog receiver, the telephone earpiece (Figure 1.11), works, using an electromagnet to convert an electrical current into a sound wave that the ear can then interpret as speech. However, if the incoming information is an electrical carrier wave that is modulated by the information signal, then the receiver must be constructed rather differently.

Just as, at the transmitter, the information is converted into an electrical signal (usually a voltage), so the analog receiver's function is to recognize that voltage and convert it back into a form suitable for use, the actual use depending on what the information is.

If the information has been modulated on to a number of carriers, the receiver's first function is to accept all the carriers. This means that its electronics have to be fast enough to allow currents to flow in response to the fastest variations in voltage on the incoming signals; in other words, the receiving bandwidth has to be as high as that of all the incoming waves. Unfortunately, this wide acceptance capability also means that it can allow in all kinds of unwanted signals and variations. These are collectively known as noise, and the noise level is a crucial feature in the reception process. Of course, the term *noise* does not necessarily mean that we can hear it with our ears, but merely that it is unwanted: there may be signals from other communications channels close by, for example, which manage to find their way in. Such "cross-talk" noise usually can be excluded by careful design, but there are also some more fundamental sources of noise, and we will now take a brief look at these.

We know that an electric current is comprised from a flow of electrons in a conductor. The magnitude of the current is, in fact, the total number of electrons that passes a given point in one second, multiplied by the electric charge on a single electron (this is a universal constant).

In any conductor, such as copper, the electrons are never still, even when there is no applied voltage. They move about randomly in the conductor, and this movement is more energetic the higher the temperature. This random movement of electrons comprises a random current, which must be added as a noise to any steady current that is present as a result of an applied voltage.

In addition to the above effect, the current resulting from a steady voltage applied to a conductor will not itself be completely steady: a careful count of the number of electrons that pass a fixed point on the conductor in a given time will not always give exactly the same answer from one measurement time interval to the next, but will vary by a small amount about an average value, as a result of this extra random component.

It is the average value that is normally quoted as the steady current flowing in the conductor as a result of the voltage applied, and that will obey Ohm's well-known law (current is proportional to voltage), but there are the above two randomly varying components also. These two types of variation in the current comprise fundamental sources of noise.

The first type of noise (random motion increasing with temperature) is called thermal noise (sometimes Johnson noise, after its discoverer), and the second type (variations in flow rate) is called shot noise (because it is similar in form to the variations in arrival rate observed when lead shot, formed by allowing molten lead to fall from a great height in a shot tower and separating into small spheres as it does so, lands at the bottom of the tower). These two types of noise add together to produce a noise level, in any electronic circuit, which can be minimized (by reducing the temperature, for example), but can never be reduced to zero. We will look at the important matter of noise in a little more detail later.

If the signal arriving at the receiver is very weak, it may be that the electric current that it produces is smaller than the receiver noise level; in other words, it is swamped by noise (Figure 1.15). In this case, the signal cannot be received satisfactorily. Any information it contains will be unintelligible, just as is normal conversation in a very noisy room. Clearly, we cannot allow the signal to fall as low as this.

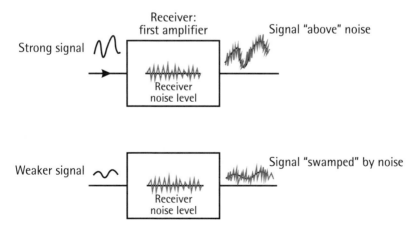

Figure 1.15 Effect of receiver noise.

The fundamental noise level caused by thermal and shot noise is a feature of the receiver itself. On the other hand, the farther away the receiver is from the transmitter, the weaker will be the received signal, owing to losses that increase with distance. In addition, the signal may be distorted: it may not have exactly the same form as the transmitted signal. Again, this will almost certainly increase with distance, so the length of the link must be chosen with this in mind also. We will discuss the reasons for losses and distortions in the transmission path in the next section.

Clearly, there will be a distance beyond which the receiver noise is larger than the received signal and/or the distortion reaches an unacceptable level. There must be a receiver before this point is reached.

Having made sure that the transmitted signal is received at a satisfactory level and in an acceptable form, it can then be increased in level by passing it through an electronic "amplifying" circuit in the receiver. This amplifier must be capable of amplifying the full range of signal frequencies entering the receiver from the channel. The amplification will boost it by a factor of 10 to 100 (depending on the requirements) so as to ensure that it is large enough for easy processing in the next stages, all of which will add some more (mostly thermal and shot) noise.

The first of these stages is frequency selection (see Figure 1.14), where the different carriers, each with their modulating signals sitting on them, are separated and passed into different parts of the circuit. Each separate signal then has its modulating signal removed. When the signal was put on to the carrier, the process was called "modulation." It is not surprising, therefore, that this removal process is called "demodulation," as we have already noted.

The lower-frequency modulation signals may then, themselves, be amplified again before passing out of the receiving circuits to their intended purposes, for example an earpiece, a loudspeaker, a TV display, or a computer.

The design and complexity of receivers varies enormously, of course. A telephone receiver looks very different from a TV set, which looks very different from a microwave receiver. But the broad principles are all the same and the receiver diagram in Figure 1.14 contains all the important elements for any analog telecommunications receiver.

1.8.2 The digital receiver

The basic function of the digital receiver is the same as for the analog receiver; that is, to accept the incoming signal from the channel and to present it in a convenient form to the outside world. The important difference is that the incoming signal is now a stream of pulses. The first function in the line is now a recognizer of pulses (or absences of pulses) in given time slots, these latter being fixed by a clock, which is itself synchronized by the pulse stream (Figure 1.16). Having recognized the pulse stream, the pulses are than demultiplexed into the various separate signals according to some pre-established scheme for the whole system, and then individually converted back into analog signals (if that is what is required) by digital-to-analog converters (DACs). The analog signals can then be separately amplified, if required.

Clearly, the recognizer circuit at the front end must, again, be able to respond fast enough to receive the incoming pulse stream and must, therefore, have an even higher bandwidth than if the signals had been in analog form. This, in turn, means even more noise than for the analog case. However, as has been noted, the act of merely recognizing a pulse (or its absence) is easier than that of establishing an accurate analog level, even in the presence of a larger amount of noise, and this almost always carries the day. The receiver has a much clearer idea of what it is looking for than in the case of an analog system. We will return to all of these important matters later, when dealing with their special importance in relation to optical systems.

So now we understand that the receiver needs a signal that is an accurate replica (of course) of the original signal and that is larger than the receiver noise.

Between the transmitter and the receiver lies the all-important communications channel. This will act to distort the signal and to reduce its size, as the signal passes along it. It is necessary now to understand something about the processes that are operating here, in order to

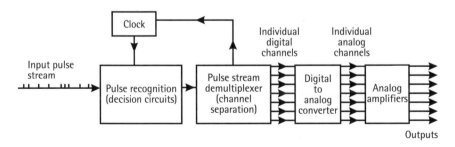

Figure 1.16 Basic digital receiver.

be aware of the crucial role the channel plays in determining system bandwidth.

1.9 Channel attenuation and distortion

An analog telephone voice signal, in electronic form, with a bandwidth of 3,000Hz (3 kHz), travels quite easily over copper wires up to distances of many tens of kilometers. However, if we seek to transmit it by modulating a carrier wave, then we know that this carrier wave must have a frequency of at least 3 kHz in order for it to contain the spread of frequencies ±3 kHz either side of the 3 kHz carrier, thus comprising a total bandwidth of 6 kHz. Hence, if we wish to transmit, say, 100 signals over the same wires (which is, as you will remember, the reason for modulating the different carriers with the signals), the minimum total bandwidth that will be needed is 100 × 6 kHz, or 600 kHz, if the various signals are not to overlap and therefore not interfere with each other (Figure 1.13c). Now, as the frequency of the transmission rises to these values, the copper wire starts to complain about the speed with which it is required to respond. The electrons within it cannot move more quickly than a certain speed and, if asked to do so, will resist. The result of this is that the electrical resistance of the copper rises. This is partly because the conduction electrons are forced into a thinner and thinner layer close to the surface of the conductor, and partly because some of the power is radiated into the surrounding space at the higher frequencies: the wire effectively becomes an antenna.

As a consequence of these effects, the higher-frequency signals progressively lose power as they travel down the wire. There will come a point, along the wire, where the higher-frequency signals are too weak to be detected satisfactorily by the receiver. The signals have suffered loss, more technically known as attenuation, too great to be tolerated. What is it that determines whether or not a signal is too weak to be received satisfactorily? We noted in the previous section that this condition is reached when the signal is weaker than all the unwanted interferences, clutter and noise that will always find their ways into any receiver system: we know that these can be minimized but they can never be reduced to zero, and their level sets a limit on the allowable channel attenuation. We cannot allow the signal level to fall below that of the receiver noise.

As we have noted, the resistance of the copper wires means that the communications signal gets weaker and weaker as it travels down the wires, suffering attenuation. We have also noted that the resistance (per unit length) is higher, the higher is the frequency (because the electrons are limited in their speed of response). Hence, the higher the frequency, the shorter is the distance over which the signal can be received satisfactorily. It follows that the greater is the required total channel bandwidth, the shorter the distance over which the total signal can be transmitted.

One way of alleviating this problem is to use amplifying repeaters. These are intermediate receiver/transmitter units placed at intervals in the path between the original transmitter and the final reception point. Their function is to receive the signal at a satisfactorily high level, increase its power level in an electronic amplifier, and retransmit it along the same path to the next repeater; and to continue this process periodically until it reaches the final destination (Figure 1.17). This is common practice on trunk telecommunications lines. However, these repeaters are expensive to make, to install in the line, and to maintain. They also add more noise and require their own local sources of power. As the frequency of the carrier rises and, hence, the copper resistance rises, it is clear that more and more repeater stations will be needed at smaller and smaller intervals. Costs will rise, and there will come a point at which the communications link becomes too expensive to be economic.

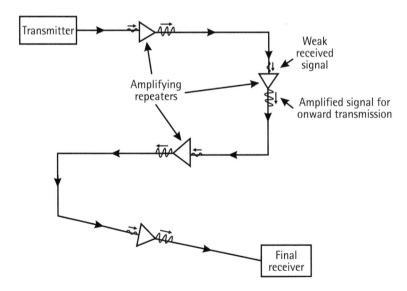

Figure 1.17 A repeater chain.

There is yet another problem. Since the resistance of the copper increases with frequency, the higher frequency components of the signal for any one modulated carrier will meet with more resistance than the lower-frequency components. This means that they will become proportionally weaker than the lower-frequency components as the signal travels down the copper conductor. The result is an unbalanced loss of signal. The higher frequencies are attenuated more than the lower ones and the signal in the channel will thus not be represented faithfully at the receiver: in other words, it will have become distorted (Figure 1.18). The copper conductor is thus a lossy medium for communications, and the loss of signal is dependent on frequency. Clearly, only a certain limited amount of distortion can be tolerated if the information present in the signal is to be adequately preserved. An amplifying repeater will not easily correct this distortion, since it will depend upon the distance traveled, and we cannot design a separate repeater for every different section length. Consequently, even in an amplifying repeater chain, the distortion will steadily accumulate.

Digitization offers a way to get around this problem, however. We saw in Section 1.6 that it is possible to digitize an analog signal; that is, to represent it as a pulse stream. Moreover, if we wish to transmit more than one analog signal over the same path, we can digitize all the signals and then interweave the pulses, perhaps taking one pulse from each stream in turn, a process known as multiplexing. Of course, the process must be reversed at the receiver so that the signals can again be separated: this is the process of demultiplexing. It was noted in Section 1.6 that one major advantage of the digitizing process is that the receiver now has only to recognize the presence or absence of a pulse in any given time slot: it does not have the more demanding task of deciding what is the level (among many, perhaps 16) of the incoming analog signals in that time slot. Hence, more attenuation can be tolerated at the receiver.

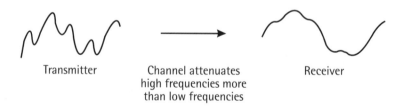

Transmitter   Channel attenuates   Receiver
                     high frequencies more
                     than low frequencies

Figure 1.18 Distortion effect of differential frequency attenuation.

The second advantage now is that more distortion also can be tolerated. The pulses may suffer increasing distortion with distance, but as long as the receiver is able to decide whether a pulse is present or absent (a binary decision) in any time slot, the distortion is irrelevant. Hence, digital systems win here too. For both these reasons, repeater distances can be increased for digital systems when compared with analog ones.

Yet another advantage is that, having properly recognized the pulse stream, the repeater can then regenerate it again for onward transmission. It is now what is called a regenerative repeater. There is now no accumulation of noise or distortion at these greater repeater distances, since the same pulse stream is accurately regenerated each time. The overall system performance has thus been improved considerably by using digitization.

This particular digital scheme, that of sampling the signal at the appropriate rate and then transmitting the sample amplitudes as a digital code in a continuous bit stream, is in fact widely used: it is known as pulse code modulation (PCM). It indeed provides the best signal reproduction at the receiver for any given received power and, of course, reduces costs, because the spacing between repeaters can be increased.

Let us tread carefully, however; there is never advantage in any engineering design without some accompanying disadvantage. What is the disadvantage of PCM? Or, as the question was posed in Section 1.6, What is the catch?

To discover this, we should refer back to the discussion in Section 1.5. There it transpired that, for a signal with analog bandwidth B Hz, with sampling at the rate of 2B samples per second and each sample being represented by an N-bit code (e.g., $N = 4$ for 16 possible levels), the implication was that the signal as a whole was represented by a bit stream at a rate of 2BN bits per second. Each bit is a pulse, and pulses at a frequency of 2BN bits per second must have a frequency up to at least 2BN Hz to represent them (see Figure 1.19). This means that the bandwidth associated with a bit stream of 2BN bits per second must be at least 2BN Hz. In other words, the original bandwidth of B has been increased by a factor of at least 2N, and this factor could easily be as high as 16 (in an 8-bit sampling system), and might well be larger. So we have traded advantages of lower acceptable signal level and lower distortion for increased bandwidth.

This is, in fact, an example of a general theorem in telecommunications (Shannon's theorem; see Appendix C for a full justification of this, if required) which tells us exactly what the relationships are between these three quantities, but which also tells us that one can only improve the signal

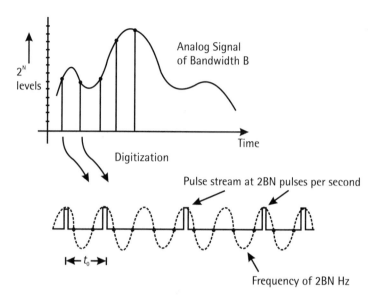

Figure 1.19 Bandwidth of a digital pulse stream.

level performance in a given channel by increasing the bandwidth. (It is for this reason that frequency modulation [FM] in radio broadcasts has a better fidelity than amplitude modulation [AM], but needs a greater bandwidth.) Of course, this requirement for greater bandwidth means that still higher channel frequencies must now be transmitted, so that the distances between repeaters must again be reduced, as a result of the high-frequency attenuation; we are back (almost) to where we were. However, if only the extra bandwidth were available, PCM would provide the most efficient scheme for using any given amount of transmitter power, owing to its remarkable noise and distortion tolerance, and this is enormously valuable, not least because it will markedly cut costs.

So, still more channel bandwidth is required and other solutions must now be sought. These solutions must allow larger bandwidths to be transmitted over greater distances. Clearly, this means that more, and therefore higher, carrier frequencies are required and that the necessary frequencies must travel with smaller attenuation and distortion than is the case with copper wires. We are seeking, in other words, to increase not just the bandwidth but also the distance over which the bandwidth can be transmitted, for it is little use having a signal with enormous bandwidth that can travel only a short distance. Thus, the requirement is to increase a quantity that is bandwidth multiplied by distance, the so-called bandwidth-distance product. (Telephone engineers have been accustomed to measure this in circuit miles, the "circuit" referring to the standard speech circuit, including its guard, of about 4 kHz.)

In order to increase this bandwidth-distance product significantly, we must look in very different directions from those that rely on the carrier being transmitted only by electrons flowing in copper wires. In the next chapter we will examine the results of these new searches for bandwidth-distance in some detail. We will see that they lead us toward optics.

1.10 Summary

Much ground has been covered in this introductory chapter. Almost all the important ideas in telecommunications have been discussed in a general and, it is hoped, understandable way. We have also become familiar with many of the words and terms commonly used in telecommunications.

This discussion was important, because it is necessary to understand why it is desirable to move to higher and higher bandwidths and higher and higher carrier frequencies thus leading, as the next chapter will show, to optical telecommunications.

We now know how to measure information, how to present it in the form of a signal, how to relate information content to bandwidth, how to modulate the signal on to a carrier wave, how to send many modulated carriers over a trunk transmission channel, and thus why higher and higher carrier frequencies are needed. We looked at some of the properties of communications channels, and noted the limited capabilities of copper wires. We compared analog and digital systems, noting the important advantages of the latter. We have looked also at noise, and at basic analog and digital receiver designs. An understanding of all of these ideas is needed for the discussion that follows.

2

Why Do We Need Optics?

2.1 The quest for bandwidth

In Chapter 1, the relationship between information content and bandwidth was explored. We saw that when information is cast into the form of a voltage signal, for example, the bandwidth occupied by the signal is greater, the greater the amount of information it represents. When this information is impressed (modulated) on to a carrier in order to send many signals over a trunk telecommunications line, the line itself has to have the capability for a large bandwidth-distance product (that is, the ability to carry a large amount of information over a great distance) if the line is to be economic as a trunk carrier.

A pair of copper wires becomes uneconomic for more than about 300 speech signals over 10 km of distance. With each speech-modulated carrier occupying about 6 kHz of bandwidth, this gives a total bandwidth of about 2 MHz and a bandwidth-distance product of about 20 MHz.km. Clearly, the conduction losses severely limit the bandwidth capabilities of copper wires and hence, for more capacity than this, it is necessary to look away from copper wires. We are looking for means of telecommunication on carrier waves at much higher frequencies, which can travel large distances with little attenuation. Around the time that this became desirable for the further development of telecommunications technology, radio waves were discovered (by Heinrich Hertz, in 1888). These waves were a particular example of a general class of waves known as electromagnetic waves. Before being able to appreciate how these have helped in the advance of telecommunications technology, it will be necessary to take a diversion to understand their primary features.

2.2 Electromagnetic waves

It had become clear during the early nineteenth century that electricity and magnetism were intimately related. Electric currents, moving electrons, were known to give rise to magnetic force fields, and moving magnetic force fields were known to give rise to electric currents.

In 1864 James Clerk Maxwell, a Scottish mathematical physicist, showed that this meant that electric and magnetic force fields ought to be able to generate each other continuously in space and thus give rise to a traveling wave: the electromagnetic wave. Maxwell went further and calculated the speed of these waves. He found this to be very close to the known value of the speed of light. It therefore became clear that light itself was an electromagnetic wave. Further experimental evidence for the nature of such waves came 24 years later, when Heinrich Hertz succeeded in producing radiating electromagnetic waves with a varying electric current in a spark chamber. Hertz was able to detect these waves by causing them to produce another spark at an unconnected point a few meters away. He called these waves radio waves, and radio was born (Figure 2.1).

The electromagnetic wave, consisting as it does of mutually sustaining electric and magnetic force fields in space, is depicted in Figure 2.2a. This wave is characterized similarly to that of the pure wave pictured in Figure 1.8a, with an additional feature that results from the fact that it now travels in distance as well as in time. If the wave is observed at one point in space, then the peaks and troughs of either field come and go with time and, as we saw in Section 1.5, the number of peaks (or troughs) that occur in one second is called the frequency of the wave; this is measured in cycles per second (Figure 2.2b), a unit that, as has already been mentioned, is given the name Hertz (abbreviated Hz). Normally, for radio waves, the symbol f is assigned to frequency.

If, now, the wave of just the electric field, say, is observed at one point in time (Figure 2.2c), we see the wave strung out in space, so that the

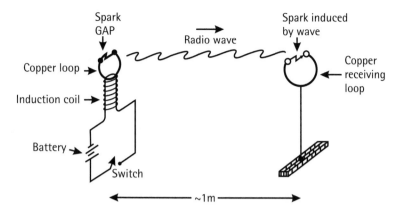

Figure 2.1 Heinrich Hertz's experiment proving the existence of radio waves (1888).

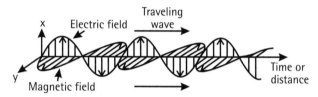

(a) Mutually sustaining electric and magnetic field variations of the electromagnetic wave

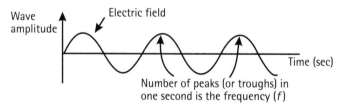

(b) The variation in time at one point in space through which the wave passes

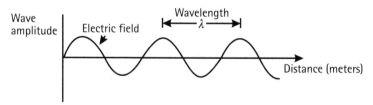

(c) The variation in space at one point in time for the traveling wave

Figure 2.2 Electromagnetic wave definitions.

wave is now also characterized by the distance between peaks (or between troughs). This distance is known as the wavelength, and it is normally assigned the Greek letter λ (lambda). The wavelength is measured in meters or, as it gets smaller, in millimeters (one-thousandth of a meter, 10^{-3} m, written as mm), micrometers (one-millionth, 10^{-6} m, μm), or nanometers (one-billionth, 10^{-9} m, nm). The wavelength comprises an important behavioral feature of the waves, as we shall soon see. It is the same for both the electric and magnetic fields, as is the frequency; in order mutually to sustain each other, the fields must vary in the same way in both space and time. Now, at one point in space through which the wave is traveling there will be f wavelengths passing in any one second, so that the speed of the wave will be $f \times \lambda$ meters per second. This is conveniently written:

$$c = f \times \lambda$$

where c is the speed of the wave. The speed of electromagnetic waves in free space does not depend on their frequency or their wavelength. It is a fundamental constant of nature. Nothing can travel faster. Its measured value is 299,792,458 meters per second and, for most telecommunications purposes, this can be well approximated by a value of 300,000,000 meters per second, or more conveniently in scientific notation, 3×10^8 m.s^{-1}. It travels only very slightly slower in air (the difference is only 0.03%), because the air offers very little resistance to the passage of the waves. Since $f \times \lambda$ is constant, it follows that as the frequency (f) rises, the wavelength (λ) must get smaller; there is no limit to the frequency (or the wavelength) that these waves can have. Hence, for example, a wave at a frequency of 3 kHz will have a wavelength of 100 km (i.e., $3 \times 10^3 \times 10^5 = 3 \times 10^8$), whereas a light wave will have a frequency of around 3×10^{14} Hz and a wavelength of around one-millionth of a meter (1 μm).

The full range of what is called the electromagnetic spectrum is shown in Figure 2.3. It extends from the low-frequency radio waves produced by Hertz's spark, for example, to the very-high-frequency gamma waves produced by some of the most violent objects and processes in the universe, such as black holes and the collapse of giant stars. With such a wide range of carrier frequencies available, it is clear that this spectrum might be of great value to us in telecommunications technology. Let us now look at how these waves can, in fact, be used in pursuit of important advances in telecommunications.

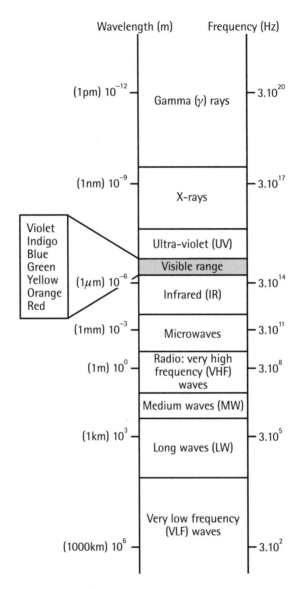

Figure 2.3 The electromagnetic spectrum.

2.3 Transmission with electromagnetic waves

Having learned something about electromagnetic waves, we next ask: What are their advantages for telecommunications? There are several.

The first is that, since the waves travel freely through air and space, no physical connection is necessary between transmitter and receiver (as Hertz discovered). This has enormous advantages, because it means that we will not be hindered by the type of resistance that appears in copper wires, notably at higher frequencies. In fact, Hertz's discovery of the electromagnetic waves known as radio waves was the first great leap forward from copper-wire-based telephony. However, it wasn't until 1894, six years after Hertz's original experiment, that Guglielmo Marconi first succeeded in transmitting information using radio waves.

The second advantage is that, as we have seen, there is now, in principle, an infinite range of frequencies available, and therefore, potentially, infinite signal bandwidth. The real position is not nearly as rosy as this, as we shall soon discover, but nevertheless, electromagnetic waves do offer the possibility for a vast increase in available bandwidth when compared with copper wires.

A third great advantage of using electromagnetic waves is that they travel very fast; we have already noted that their speed is very close to 300,000,000 meters per second; nothing can travel faster than this (this is one of the cornerstones of Einstein's special theory of relativity). It means, of course, that there is very little delay in receiving the transmitted messages: electromagnetic waves can travel all the way around the circumference of the earth in only about one-seventh of a second.

How, then, should we use these waves in our trunk telecommunications channels? To answer this, let us first examine how such waves can be generated. We need to generate waves of electric and magnetic force fields, and this can be done conveniently by forcing electrons to move rapidly back and forth along conducting wires, since pressure of electrons, when they are all compressed at one end of a wire, creates an electric force, and moving electrons, an electric current, creates a magnetic force. These force fields move away from the wires as a pair of mutually sustaining waves to form the electromagnetic wave. These generating wires are then called "antennas." Hertz's spark was effectively an antenna, since it involved rapidly oscillating electrons (albeit now not in a wire), as do all sparks. The problem for telecommunications is that such antennas radiate waves in all directions, rather than just toward a particular receiver along a trunk line (Figure 2.4). This means that any one receiver will receive only a small fraction of the transmitted power, and the farther away it is, the less power it will receive. This wave spreading thus comprises, effectively, a source of attenuation. Of course, it is sometimes very convenient for a transmitter to

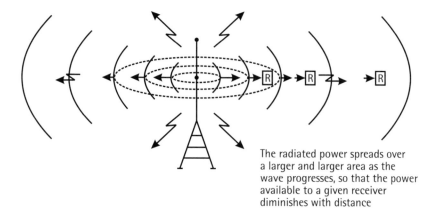

The radiated power spreads over a larger and larger area as the wave progresses, so that the power available to a given receiver diminishes with distance

Figure 2.4 Uniformly-radiating antenna.

be received by a large number of receivers spread over a wide area, and this is what happens for national radio and television broadcasts; it is also convenient when the receiver is moving around an area, as in mobile telephony. But these are relatively narrow-bandwidth applications. The rates of information transmission are quite small. For very-wide-band trunk applications, point-to-point links are required. So how can this spreading of the transmitted power be overcome?

One solution is to guide the waves in some kind of conducting arrangement such as is shown in Figure 2.5. Here there is a central conducting wire of copper lying along the axis of an outer copper cylinder, with the space between the two elements filled with an insulator, such as polythene. This is known as a coaxial cable (since both wire and cylinder share the same long axis) and it comprises a flexible guide for electromagnetic waves.

Since conductors are involved in this arrangement, some of the wave power depends on the movement of electrons in the conductors, and it may seem that we are saddling ourselves with the same attenuation problems as in the case of copper wires. However, in this case, most of the power is carried in the electromagnetic wave between the conductors, and only a relatively small amount in the flow of electrons (Figure 2.5). The result of this is that very much higher wave frequencies can be used before the attenuation becomes too severe, and the closed structure prevents the copper wire from itself becoming an antenna, which would cause it to radiate signal power away into the surrounding space (this feature was mentioned as a source of attenuation in Section 1.9). Taking a minimum economic

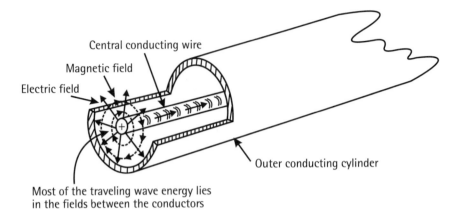

Central conducting wire

Magnetic field

Electric field

Outer conducting cylinder

Most of the traveling wave energy lies
in the fields between the conductors

Figure 2.5 The coaxial cable.

repeater spacing of about 20 km, we find that it is possible to transmit around 3,000 speech signals simultaneously over a coaxial cable, corresponding to a bandwidth of about 20 MHz and a bandwidth-distance product of 400 MHz.km. This is about 20 times more than was possible with a pair of copper wires.

Much more bandwidth is still needed, however. So far we have been concentrating on voice channels, but there is much more to telecommunications than just these. There is facsimile (still pictures), video (moving pictures), and computer data, for example. For a video signal (e.g., television) about 5 MHz of bandwidth is required, and for computer data just about as much as possible, in the longer term, anyway. Hence we must look toward higher and higher carrier frequencies. It is clear that the electrical resistance of copper (or any conductor) will eventually always impose a severe limitation on bandwidth, whether it is carrying a current or guiding a wave, so we must look beyond that. The obvious first direction in which to look is toward the free-path transmission of high-frequency electromagnetic waves through the atmosphere. There are three questions that arise immediately in regard to this: First, how can such waves be directed in narrow beams so that point-to-point links can be constructed for trunk lines? Second, how well do such waves travel through the atmosphere? And third, how easy is it to generate such waves? Let us look at these three questions in turn.

First, how can we produce narrow, directed beams of electromagnetic waves? Well, we saw earlier that one way of producing electromagnetic waves is to cause electrons to oscillate rapidly back and forth in a conducting wire known as an antenna. The ideal length of wire needed to generate waves of a particular frequency is equal to about half a wavelength at this frequency. The wire is said to resonate, like a plucked violin string, at this frequency when it has the correct length (Figure 2.6b), just as the violin string will generate sound waves with a half-wavelength equal to its length (Figure 2.6a).

If we are going to produce narrow beams of such waves, it is necessary for us to design structures that are geometrical arrays of these half-wave antennas (known as dipoles) with half-wavelength spacings, so that the waves from the various antennas can reinforce in the required direction and cancel in other directions, by a process known as wave interference. (We will look more closely at this process in Chapter 3). Constructing such arrays is much easier when the wavelengths are small and thus when the antennas are short. For example, medium-wave radio broadcasts, at frequencies of around 1 MHz, operate at wavelengths of about 300 meters (i.e., 3×10^8 m.s^{-1} ÷ 10^6 (Hz) = 300 m), so that very tall masts are required, of order 150 meters high (which is half this wavelength). Clearly, arrays of masts with these heights and spacings are quite impractical. It is much easier to direct the waves when the wavelengths come down to about 1 meter. This corresponds to a frequency of 300 MHz, entering what is known as the

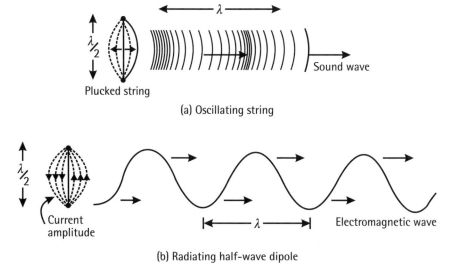

(a) Oscillating string

(b) Radiating half-wave dipole

Figure 2.6 The half-wave radiator.

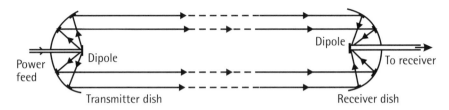

(a) Narrow beam communication with microwave dishes

(b) A microwave dish for satellite reception

Figure 2.7   Microwave dishes. (BT Corporate Picture Library: A BT Photograph.)

"microwave" range (see Figure 2.2a). The added advantage of moving to the higher frequency is, of course, that it also offers greater bandwidth.

Microwave antennas can be quite compact and, clearly, will become more so as the frequency rises. The most effective kind of microwave antenna is the well-known microwave "dish," such as is used for receiving satellite television. This has a diameter of about a meter. The dish is made of conducting material, and the wavelength of the radiation is smaller than the dish diameter. A dipole is placed at the focus of the dish (see Figure 2.7a) and, when fed with electrical power, it radiates in all directions. Wherever radiated waves from the dipole strike the conducting dish, they effectively create secondary, radiating dipoles. Thus all of these dipoles, on the inner surface of the dish, form the equivalent of an ordered array that, owing to the phases of the waves resulting from the geometry, all reinforce in just one direction: the forward direction. Hence the radiation is narrowly collimated in this direction, toward the receiving dish (Figure 2.7a). When it arrives at the receiving dish, the same process happens in reverse, and the incoming radiation is focused on to a dipole placed at this dish's focus. This results in an electric current that then passes to the receiving electronics. A photograph of a larger microwave dish for receiving satellite transmissions is shown in Figure 2.7b.

With such dishes, microwave beams can be transmitted effectively over distances of up to 250 km before beam spreading takes over again and significantly attenuates the signal at the receiver. So now we have the answer to the first question: that of how to produce narrow, directed beams of radiation.

The second question concerns how easy it is to generate such waves. We know that we must make electrons oscillate in a conductor, but to do this at these very high frequencies means that the electrons have to move very fast and hence have to be fed with a lot of energy, and in such a way as to force them to oscillate back and forth at these very high rates. How can this be done?

The problem was first solved satisfactorily during World War II, when the necessity for effective radar systems at microwave frequencies motivated a lot of effort to develop high-power sources of microwaves. Radar played an important part in the defenses of southern England during the Battle of Britain in 1940, and its success was largely due to microwave sources known as the klystron and the magnetron. These devices provided sources of high power at wavelengths of about 10 cm (corresponding to frequencies of about 3,000 MHz). The magnetron is also used today in microwave ovens. There have been other important developments since.

So we now have both the sources and the directional arrays necessary for trunk-line, free-path transmission at frequencies up to 3,000 MHz and beyond, giving us the potential for another large increase in trunk bandwidth-distance. Unfortunately, however, not all of this 3,000 MHz carrier is available as signal bandwidth. The main reason for this is that any given microwave source cannot provide power over all the frequencies up to 3,000 MHz. It is essentially a high-frequency device, and it provides energy within a fairly narrow range of frequencies, of order 5% of the central carrier. This gives a bandwidth of around 150 MHz at 3,000 MHz. Hence we can modulate the carrier, either with the analog signal or the equivalent digital pulse stream, up to this rate. At first sight this looks little better than the coaxial cable, which had a bandwidth of about 30 MHz; but only over a distance of about 20 km. Our new microwave bandwidth of 150 MHz can travel a distance of around 250 km in the atmosphere. This gives it a bandwidth-distance product of 37,500 MHz.km, an increase of almost 100 times over that for a coaxial cable.

It is clear that the frequency units have again become cumbersome, so that yet another adjustment is needed. It is convenient now to define a frequency unit equal to 1,000 MHz: the gigahertz (10^9 Hz), abbreviated GHz. The bandwidth-distance product above now becomes 37.5 GHz.km.

Still more bandwidth-distance is needed, so higher and higher we go. There are sources available at 10 GHz (10^{10} Hz) and thus the bandwidth available from these is of order 500 MHz. This is enough to handle 100,000 speech signals or 100 video signals, and again the radiation can be beamed over 250 km or so with the aid of a microwave dish.

Clearly, these beamed microwave signals need a direct line of sight between transmitter and receiver. Microwaves cannot bend around the earth or around buildings, as the much longer wavelengths can do. Roughly speaking, waves can only bend around objects that have a size smaller than their wavelength. A wave can, effectively, curl itself around an object smaller than the rate at which it, itself, is changing in space. At 1 GHz the wavelength is only 30 cm, so that most man-made objects in towns or cities, being larger than this, will block them. It is for this reason that many larger cities have now constructed tall telecommunications towers, such as the BT Tower in London (Figure 2.8). These towers are high, slim structures that support sets of microwave dishes close to their tops, in order to allow direct line of sight transmission paths over the tops of the city buildings to, perhaps, another such tower in another city.

Satellite communications also use frequencies in this range (1 GHz to 10 GHz). An artificial satellite is placed in an orbit where it revolves at the same rate as the earth, a so-called geostationary orbit (that is, at a height of about 35,000 km). It thus remains fixed with respect to the earth's surface and it can be used to "bounce" microwaves between any two points on the earth to which it is visible (Figure 2.9). Clearly, as the waves are going almost straight up and down, there is little possibility of obstruction. One difficulty, however, is that, because of the large return distance, there is now an appreciable time delay, of about one-third of a second, between transmission and reception. This can be troublesome, and irritating, during a two-way conversation.

So now we have a formidable range of carrier frequencies for wide-band telecommunications. These are summarized in Table 2.1. They range from the very low frequencies (VLF) carried by copper wires through the

Figure 2.8 The BT telecommunications tower in London. (BT Corporate Picture Library: A BT Photograph.)

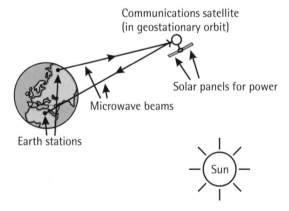

Figure 2.9 Satellite communication.

Table 2.1   The Usage of the Radio Spectrum

Name of Band	Frequency Range	Usage
Very low frequency (VLF)	3–30 kHz	Very long distance communications
Low frequency (LF)	30–300 kHz	National broadcasting Radio navigation
Medium frequency (MF)	0.3–3 MHz	National broadcasting
High frequency (HF)	3–30 MHz	Radio telephony
Very high frequency (VHF)	30–300 MHz	FM broadcasting Television broadcasting Mobile radio telephony Radio navigation
Ultra-high frequency (UHF)	0.3–3 GHz	Television broadcasting Mobile radio telephony Radio navigation Radar
Super-high frequency (SHF)	3–30 GHz	Multi-channel trunk telephony Radar Satellite communications
Extra-high frequency (EHF)	30–300 GHz	TE_{01} waveguide (?)

low (LF), medium (MF), and high (HF) frequencies that can be carried by coaxial cable, up to the VHF and microwave frequencies, which can only be transmitted effectively through the atmosphere and free space.

But the demand for more and more bandwidth is relentless. Higher and higher we must go in our search to meet the ever increasing demands of the computer age. However, at around 10 GHz we encounter another major obstacle. At this frequency the atmosphere begins to absorb, and therefore attenuate, the microwaves quite significantly, especially when it contains moisture in the form of mist, fog, or rain (see Figure 2.10). The trunk telecommunications network, as presently configured (2000), makes good use of microwaves up to about 10 GHz, but what is to be done to acquire even greater bandwidth in the face of this atmospheric blockage?

During the 1970s there was an attempt to guide waves with frequencies up to about 100 GHz (i.e., wavelengths in the millimeter range) in hollow conducting tubes (no central conductors in this case), known as cylindrical waveguides. The idea here was that these tubes could be pumped free of air and moisture and that, therefore, the attenuation would be greatly reduced

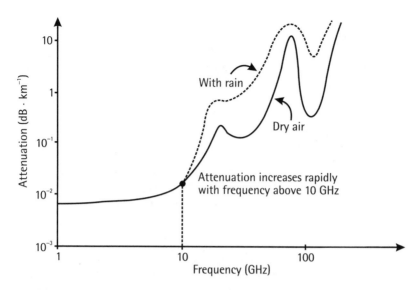

Figure 2.10 The variation of microwave attenuation with frequency for two types of atmospheric condition.

(Figure 2.11). Waveguides for microwaves were not a new idea. They had been used for some time, with either square or rectangular cross-sections, to guide microwaves over short distances (meters), for convenience in taking power to transmitters or from receivers, for example. But the copper conduction losses (the losses discussed in Section 1.7) in them were too severe for use in long-distance transmission at these frequencies. However, the idea that arose in about 1970 was that of using a particular traveling wave pattern, known to the experts as the TE_{01} mode, in which almost all the wave power was carried in the electromagnetic fields, and almost none in the conduction electrons. Consequently, the attenuation was very small indeed; but only when the waveguide was straight. These waveguides were actively researched during the 1970s, but they were temperamental, needed expensive precision engineering, and careful (almost) straight-line installation. Fortunately, other developments were afoot that were soon to overtake the TE_{01} waveguide.

These developments were the result of research, also during the 1970s, on an entirely new type of waveguide. This waveguide contained no metallic conductors at all, and it guided not microwaves but light waves. It was called the optical fiber.

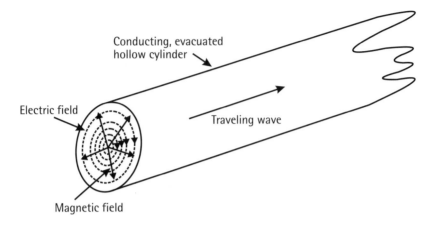

Figure 2.11 High-frequency microwave waveguide (TE_{01} mode).

As we know, light is a form of electromagnetic wave, so the only actual difference between the light waves and microwaves is that of the wave frequency. The optical fiber guides light waves with frequencies of around 300,000 GHz. Clearly, we now need yet another frequency unit: the terahertz (10^{12} Hz), abbreviated THz. So now we are dealing with frequencies of the order of 300 THz, and wavelengths of the order of one-millionth of a meter, or one micrometer (1 μm). This wave frequency lies within what we call the infrared region (see the spectrum, Figure 2.3), which is only just beyond the range of human vision. It is generally classed as "light," since it obeys the laws of optics (which are to be covered in the next chapter) and can, for example, be focused by glass lenses, reflected by glass mirrors, et cetera. The collimation and focusing of microwaves was beginning to show such behavior, as we noted when discussing them (see again, for example, Figure 2.7a). The changes in wave behavior are gradual as we pass along the electromagnetic spectrum and, in any case, are changes of scale rather than of kind.

The range of human vision (i.e., the range to which the human eye's retina is sensitive) is from a wavelength of 0.7 μm, which corresponds to red light, down to 0.4 μm, which corresponds to violet light. Hence, just above 0.7 μm we speak of infrared light and just below 0.4 μm, of ultraviolet light. Between 0.7 μm and 0.4 μm we have the full visual color spectrum: red, orange, yellow, green, blue, indigo, violet. Green light, in the middle of the range, has a wavelength of about 0.5 μm and a frequency of about 600 THz.

For convenience, the notation for all the frequency and wavelength units discussed so far is summarized in Tables 2.2 and 2.3.

Since this new optical waveguide is not made from metal, it involves no conduction electrons and thus no electrical losses. It therefore comprises a very-high-frequency (around 3×10^{14} Hz) carrier system with very low loss. This was just what the telecommunications industry needed. Almost overnight, it rendered the TE_{01} waveguide redundant.

The bandwidth offered by an optical wave, taking 5% of 300 THz, is of order 15 THz, or 15,000 GHz. This is enough for every member of the human race to be talking to another member at the same time—and all along one optical-fiber waveguide. Again, we may modulate the optical carrier, either with the analog signal or, better, with the digital version, in which case we are dealing with a stream of light pulses. Clearly, this optical regime has the potential to meet all our telecommunications bandwidth requirements well into the foreseeable future, possibly for the whole of the twenty-first century. Additionally, the optical fiber is very thin (about 100 μm in diameter), is of low weight, is easily bent around corners, and is made from sand, one of the cheapest and most abundant materials on our planet. This is why a global optical-fiber telecommunications network is presently under construction.

TABLE 2.2   Multiple Units

FREQUENCY (HZ)	SCIENTIFIC NOTATION (HZ)	NAME	ABBREVIATION
1,000	10^3	Kilohertz	kHz
1,000,000	10^6	Megahertz	MHz
1,000,000.000	10^9	Gigahertz	GHz
1,000,000,000,000	10^{12}	Terahertz	THz

TABLE 2.3   SUB Multiple Units

WAVELENGTH (FRACTIONS OF A METER)	SCIENTIFIC NOTATION	NAME	ABBREVIATION
Thousandth	10^{-3} m	millimeter	mm
Millionth	10^{-6} m	micrometer	μm
Thousand millionth	10^{-9} m	nanometer	nm
Million millionth	10^{-12} m	picometer	pm

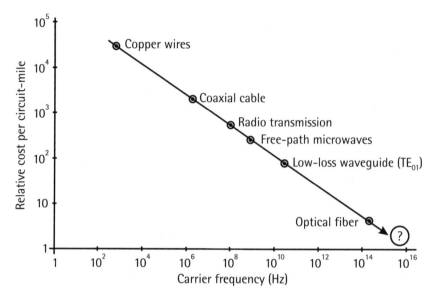

Figure 2.12 The economics of increasing carrier frequency.

The economic advantages of being able to pass more and more signal bandwidth down one trunk-line channel are illustrated in Figure 2.12. The diagram shows how dramatically the real (i.e., inflation-adjusted) cost for unit bandwidth-distance product, measured in units of kHz.km, has fallen with increase in carrier frequency, as the changes in technology have occurred. Already, for example, optical-fiber technology has reduced costs by a factor of about 10,000 compared with the early days of copper conductors, and this new technology is only in its infancy. A reduction by another factor of 10 certainly is soon to be achieved and, in the medium term, probably a factor of 100. With such enormous bandwidth capability available at such low cost, there is the near certainty of another major impact on the whole structure of society: on the way we work, play, learn, and generally live our lives.

Now that we have gained an appreciation of the developments that have led to optical-fiber technology and of the position it occupies in the overall scheme of telecommunications, it is time to begin to understand how optical fibers and optical-fiber telecommunications systems actually work. This is the purpose of the rest of the book. Our first task must be to become familiar with the optical fiber itself. This is the subject of the next chapter.

2.4 Summary

In this chapter we have seen how the relentless need to increase transmission signal bandwidth or, more specifically, bandwidth-distance, leads us away from copper wires as the carriers of information, to electromagnetic waves. We studied the basic properties of these waves and then looked at the ways in which these waves can be used in trunk-line telecommunications networks. This led us first to the coaxial cable, on to beamed microwaves through the atmosphere, and then, because of the attenuation by moist air above about 10 GHz, to the evacuated microwave waveguide.

We finally noted that the above systems are being superseded by an entirely new type of waveguide: the optical fiber. This waveguide guides light waves, and it contains no conducting material at all. There are, therefore, no losses of the type caused by electrons flowing in conductors, and the losses are very low indeed in this waveguide. This, coupled with the fact that optical frequencies are very high, at around 300 THz (3×10^{14} Hz), means that optical-fiber technology offers the means for providing all of mankind's telecommunications bandwidth requirements well into the foreseeable future, perhaps for the whole of the twenty-first century.

3

What Are Optical Fibers?

3.1 Introduction

We saw in Chapter 2 that a light wave can act as a carrier of large quantities of information; that is, as a high-bandwidth carrier. We also saw that, in order to use this advantage in a large-capacity communications system, it is necessary to guide the light in a low-loss medium: the optical fiber.

We shall now look more closely at the construction and properties of optical fibers, and the way in which they guide light. This will enable us to understand the problems that had to be overcome before today's operational networks could work as well as they do, and those that are still to be overcome if higher and higher information transmission rates are to be achieved.

3.2 What is an optical fiber?

An optical fiber is a thin wire of glass. Its diameter is about the same as that of a human hair, approximately one-tenth of a millimeter (0.1 mm). Light can be guided in such a wire by launching it in to one end, using focused light from an intense source, and allowing it to bounce down to the other end by a series of reflections from the sides (Figure 3.1). A similar effect is sometimes used in public fountains for display purposes: jets of water are fed with light of various colors at one end, and the light is seen to be guided along the jets to provide a pleasing, varicolored, sparkling effect. Quite a lot of light escapes from the jets as it progresses, however (this is why it is such an impressive, sparkling display), so that, from the telecommunications point of view, it represents an extremely lossy medium. In order to reduce this loss, it is necessary to understand better the basic processes involved.

Light travels more slowly in glass than in air or in a vacuum. The reason for this is that the light wave frequently meets the densely packed molecules of which the glass is composed and hence is impeded by them. This interaction between the wave and the glass material is a fairly complex affair: it depends on the properties of the wave and the properties of the molecules, and the guiding of light within glass depends crucially on a number of these properties, as we shall soon see. However, for the moment, let us concentrate on the effect that they have on the speed of the wave.

When a light wave passes from one material into another where its speed is different, it changes direction. A well-known, everyday manifestation of this is the apparent bending of a stick when it is pushed vertically into water (Figure 3.2): the speed of light in water also is less than that in air, since its molecules are, as in glass, more densely packed than in air. The reason for the apparent bending is illustrated in Figure 3.3a. Since the speed of the wave is greater in the air than in the glass, the distance traveled by the wave in going from A to B in the air takes the same time as that taken in traveling from C to D in the glass. Hence, as shown in the diagram, the wave slews around and its direction is changed. This is the effect known as refraction.

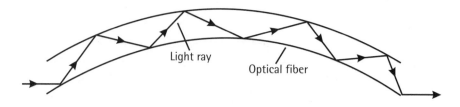

Figure 3.1   Ray bouncing inside an optical fiber.

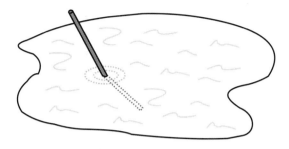

Figure 3.2 Apparent bending of immersed object in water, owing to refraction.

The slowing down of light in any given material is characterized by means of its refractive index. This is a number that represents the factor by which the light is slowed in the material when compared with its speed in free space. It defines a kind of optical density. For example, the refractive index of glass is about 1.5 (it varies somewhat depending on the type of glass and the optical wavelength), so that the speed of light in glass is about 1/1.5, or two-thirds of its value in free space. The refractive index of water is about 1.33.

Not all of the light is refracted when it moves from one medium to the next, however. Some of it is reflected (Figure 3.3a, ray E to F), the strength of the reflection depending on the densities of the two materials and the angles of the rays. It is for this reason that we can see ourselves reflected in windows, especially when the far side is in darkness so that there is no refracted light coming to us from that side to distract us. If, now, we concentrate on just one ray (Figure 3.3b) the refractive bending can be well represented by noting its change of direction as it passes from air, a less dense medium, to glass, a more dense one. Of course, the times of passage (A to B and C to D) are just the same in the opposite directions (B to A and D to C), so that the same diagram, with oppositely directed arrows, would represent the passage of the light from glass to air (Figure 3.3c). Note that, when passing from glass to air, the light ray in the air bends more toward the boundary between the media; that is, angle i is greater than angle r. (The reverse is, of course, true when the ray passes from air to glass).

Consider now what happens as angle i is decreased. Angle r also decreases and the refracted ray gets closer and closer to the boundary until (Figure 3.3d) it comes to lie along the boundary. At this angle, and for angles i that are smaller than this, there can be no refracted ray; all the light is reflected. The angle at which the light is reflected is also i (Figure 3.3d).

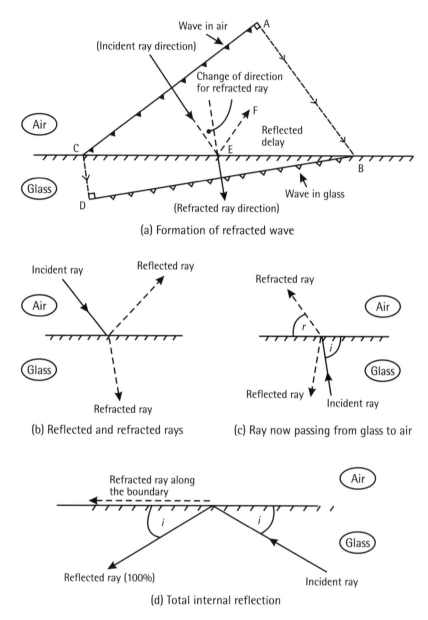

Figure 3.3 Ray optics at a glass/air boundary.

This comprises the law of reflection, and it is just as true for (non-spinning) billiard balls against cushions as for light rays. This is the phenomenon known as total internal reflection (TIR), and it is vitally important for the understanding of how light is guided in an optical fiber.

One of the many everyday uses of TIR is in efficient reflection for optical devices such as cameras and binoculars. The prism reflector (Figure 3.4) has the advantage of a reflection at a surface that makes no use of metal films as normally used in metallized hand or body mirrors. The metal will peel and corrode with time; the glass only needs to be kept clean, for normal usage.

The fact that there is a definite angle at which TIR first occurs when light passes from a more dense to a less dense medium is the reason that light is guided by a jet of water. To see the action in glass, suppose that we consider a slab of glass in air, such as in Figure 3.5. If a ray, already inside the slab, strikes the top side of the slab at an angle i smaller than that needed for TIR, then, since the angle of reflection also will be equal to i, this will also be the angle at which it strikes the bottom side, so that TIR will again take place there. The net result is that the light ray will bounce down the slab to emerge at the far end: the light has been "guided" from face A to face B by the slab. Similarly, light is guided from one end of a water jet to the other.

Let us look now at a cylinder of glass with the structure shown in Figure 3.6. Here there is an inner cylinder of glass, which we will call the core, surrounded by an outer cylinder of glass of slightly smaller density, which we will call the cladding. From the ideas we have been discussing above, it is clear that a ray of light in the core, if it lies at a sufficiently

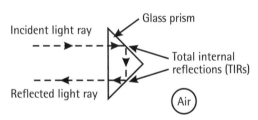

Figure 3.4 Use of TIR as an ideal mirror.

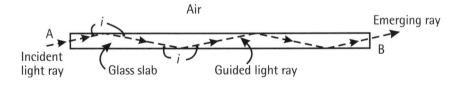

Figure 3.5 Guiding of light in a glass slab.

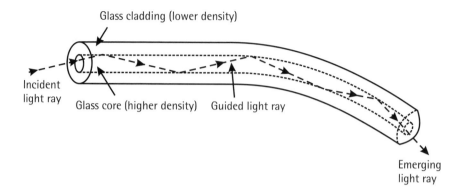

Figure 3.6 Guiding of light in an optical fiber.

shallow angle to the boundary between the two types of glass (i.e., the core/cladding boundary), will experience TIR, and the light will bounce down inside the core from one end of the cylinder to the other. The light is now guided by the glass core. This is the action of a glass fiber: the optical fiber.

Let us consider some of the advantages of using such a structure to guide light.

First, by using the two-glasses, core/cladding structure rather than a glass/air one, we ensure that the boundary at which TIR takes place is entirely protected from the atmosphere. It cannot, for example, get dirty or wet; if it did, the TIR condition would be changed and the guiding action would be disturbed.

Second, the cylinder can be very thin—just a glass wire, in fact—so that it can be flexible and easily handled. A typical optical fiber, as was stated earlier, has a cladding diameter of about one-tenth of a millimeter, which is about the same as a human hair, and it is almost as flexible. Furthermore, the light can still be guided when the fiber is bent, provided that the bending is not so severe that the ray is forced to fall outside the TIR condition when it strikes the boundary. We can do our best to avoid this by making sure that we are well within the TIR condition in the first place; that is, that i is rather smaller than for when TIR first occurs. If the bending is still too severe and the ray does fall outside the condition, it will lose some of its power to the cladding via a refracted ray, in much the same way as light is lost from the sides of the water jet. We shall examine this effect in more detail later, but it is clear that this ability to bend the fiber (not too severely) means that we can use it to take light from any given point to any other point via almost whatever path we choose. Clearly, this is going to be very useful in a telecommunications system.

Third, the glass can be made very pure using well-developed manufacturing techniques, so that light can pass along the fiber with very little loss. This means, of course, that it can be used as a telecommunications medium. However, in order to see how good a telecommunications medium it is, it is necessary to look in rather more detail at just how the light interacts both with itself and with the glass as it travels down the fiber.

3.3 Wave interference

We know from the discussion in Chapter 2 that light is a form of electromagnetic wave. An important property of all waves is that they can interfere with each other. Examples of wave interference occur in everyday life, one of the best-known being the interference of waves on water. If a stone is thrown into a still pond, the waves will be seen to move out as circles of the rising (peaks) and falling (troughs) of the surface of the water, centered on the point of the stone's entry into the water. If another stone of similar size is thrown into a different point while the first stone's waves are still present, the two sets of waves will be seen to interfere with each other (Figure 3.7). For example, if a trough of one wave happens to coincide with the peak of the other, the two cancel each other out, and the water is left undisturbed (the "spokes" in Figure 3.7). On the other hand, if two peaks (or troughs) coincide, the water level is then twice as high (low) as before. At other points on the pond's surface, the level lies somewhere between these two extremes.

Exactly the same effect occurs with electromagnetic waves, including light waves. An everyday example of this is the colorful effect one sees in an oil film on water in the road. These colors result from the fact that light from the sun is reflected from both the top surface and the lower surface of the film, which is in fact very thin, of the order of the wavelength of light (approximately 1 μm). If the peaks in the upper reflected wave coincide with those in the lower reflected wave for a particular optical wavelength (Figure 3.8), then that wavelength is reinforced and the corresponding color is bright; for a different color a peak may coincide with a trough. Cancellation will then occur and that color will be absent. As the observer moves his or her head, the path lengths change, and so the pattern of colors changes also (Figure 3.8).

Figure 3.7 Interference of water waves.

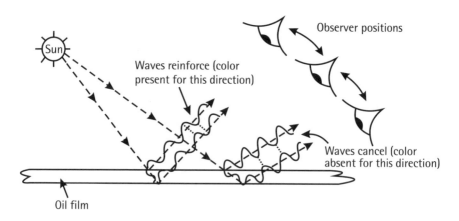

Figure 3.8 Oil film colors due to wave interference.

Let us look now at the application of these ideas to the light waves bouncing down an optical fiber. The various waves at the different angles (all smaller than that required for TIR) will interfere with each other, in a rather complicated way, to form complex interference patterns of wave amplitude. However, for certain angles defined by relationships between the geometry of the fiber and the optical wavelength (as in the case of the oil film), the same pattern will reproduce itself (the condition for this is that all waves traveling in the same direction must be in step; that is, peaks and troughs must coincide) continuously along the fiber (see Figure 3.9a). All other patterns will be changing continuously and will eventually die away to zero amplitude. The patterns that do self-reproduce are known

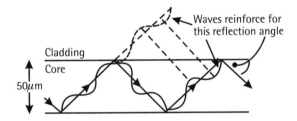

(a) Reflecting waves all reinforce for the allowable waveguide modes

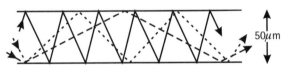

Relatively thick fiber allows many reinforcing angles within the TIR condition. This is a "multimode" fiber. Each mode progresses at a different speed.

(b) Multimode fiber

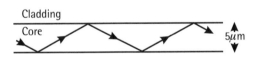

Thin fiber allows only one "reinforcing" angle within the TIR condition. This is a "monomode" fiber. The one mode means only one mode speed.

(c) Monomode fiber

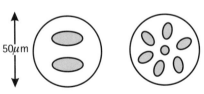

Two multimode interference patterns of light emerging from the fiber end.

Pattern of light emerging from a monomode fiber.

(d) Some interference patterns (modes) for the two types of fiber

Figure 3.9 Formation of modes by wave interference in optical fibers.

as the propagation modes of the waveguide, because they comprise just the vehicles by which light can pass from one end of the fiber to the other (Figure 3.9a).

Since each mode corresponds to a particular ray angle, it is clear that the rays for the different modes will progress down the fiber at different speeds (Figure 3.9b). This is an unsatisfactory situation for telecommunications use, because it would mean that the information that was being carried by the light would have a variety of arrival rates at the receiver. This would, in a digital (i.e., pulsed) system, for example, cause the pulses to tend to spread into each other as they progress along the fiber (as will be shown later in Figure 3.14), and thus limit the rate (the bit rate) at which they could be sent. In other words, it would limit the bandwidth. Consequently, the fiber geometry is arranged, in relation to the optical wavelength being used, so that only one mode can pass down the guide; that is, there is only one ray angle, within the TIR condition, that can cause a self-reproducing pattern. Such a guide is called a monomode (or single-mode) fiber (Figure 3.9c), to distinguish it from the multimode fibers that allow more than one mode to progress. In this type of fiber, the information travels at just one mode speed, and hence arrives in a compact state at the receiver. For this reason we always choose to use monomode optical fiber for long-distance telecommunications (although multimode fiber is still sometimes used for distances of a few hundred meters or so) in order to avoid this modal limitation on bandwidth.

The monomode fiber needs a core of small diameter, about 5μm (five-millionths of a meter), and a cladding whose refractive index is only about 1% less than that of the core. Multimode fibers have core diameters up to about 50μm. Figure 3.9d gives examples of the light patterns emerging from the fiber end for the two types of fiber.

But what about the glass itself? How does the glass affect the light passing along the fiber, and what effects does the glass have on the information transmission process in a telecommunications system? To answer these questions we must look more closely at the way in which the optical wave interacts with the glassy material of the fiber.

3.4 Attenuation and dispersion

When light passes through glass (or any other material), there is some interaction between the light and the glass. There are two main consequences of this interaction. The first is that some of the light power is lost, thus attenuating the optical signal. The second is that there is a change in the relative timing of the various optical frequency components, leading to a signal distortion. This latter effect is known as dispersion.

Clearly, both of these effects are going to be important in the design of optical telecommunications systems.

3.4.1 Optical attenuation

The glass used in optical fibers is a very pure form of amorphous (i.e., noncrystalline) silica—in other words, a pure form of fused sand. It consists, therefore, almost entirely of molecules of silica, each one of which contains one atom of the element silicon and two atoms of the element oxygen. The amorphous nature of the material means that the silica molecules are arranged randomly, not at all in an ordered framework such as would be found in crystalline quartz, which is another form of silica. In this disordered structure we find (when examined under an electron microscope, for example) clumps of molecules, of various sizes and shapes.

When a light wave passes through the material, it encounters this clumpy structure. A wave that has a small wavelength (blue end of the spectrum, say) finds it very difficult to thread its way through these clumps of molecules (Figure 3.10a) so that it is scattered sideways and backwards, and much of it is lost from the forward-progressing light path. As the wavelength gets longer, the wave can more easily thread its way forward (Figure 3.10b) and it is scattered less.

This scattering effect is thus strongly dependent on the wavelength: in fact, it decreases as the fourth power of the wavelength so that when the wavelength doubles ($\times 2$), for example, the scattering is sixteen times ($\times 2^4$) less strong. The effect is known as Rayleigh scattering (see Figure 3.10c) and, clearly, it removes power from the forward-traveling light: some of it goes back toward the source; other portions of it hit the core/cladding boundary outside the TIR angle condition and will refract into the cladding, no longer being guided. Rayleigh scattering thus comprises a severe source of attenuation for the telecommunications signal at the shorter wavelengths of light.

Rayleigh scattering is also responsible for the blue color of the sky. The reason for the blue color, when looking away from the sun, is that blue light, having a shorter wavelength, is scattered sideways much more strongly than is red light, of longer wavelength (see Figure 3.10d). In fact, since blue light's wavelength is only about half as great as that of red light, it

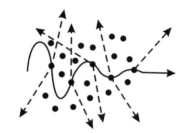

(a) Short wavelength is easily scattered by the medium

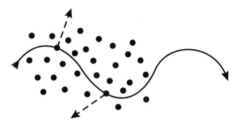

(b) Longer wavelength threads its way between the molecules

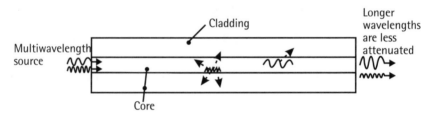

(c) "Rayleigh scattering" in optical fiber as a result of (a) and (b)

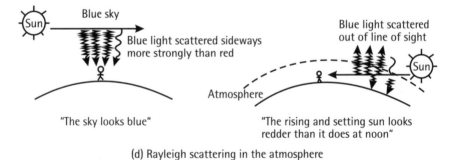

(d) Rayleigh scattering in the atmosphere

Figure 3.10   Dependence of material scatter on the wavelength of light.

is scattered about 16 times as much, as was noted earlier. It is for the same reason that the sun looks red (the blue light has now been preferentially scattered out of the line of sight) and even redder at sunrise and sunset, when the light has to travel through a greater thickness of atmosphere to reach our eyes (see Figure 3.10d), and suffers more of this scattering as a result.

As the wavelength continues to increase, the scattering in the silica continues to fall, but another effect kicks in. The silica molecule has the kind of structure shown in Figure 3.11a, with the oxygen atoms attached to the silicon atom by chemical bonds. This structure can vibrate in a variety of ways, two of which are illustrated in Figure 3.11b. The frequencies at which these vibrations occur depend on the strengths of the chemical bonds and the weights of the atoms, but the frequencies are of the same order as those of the longer wavelengths of light (remember that speed of light = wavelength × frequency). This means that these longer wavelengths set the molecule into vibration and, when this happens, the energy of the vibration is taken from the light wave itself. This, therefore, is another source of loss of power from the propagating wave. The wave, at these

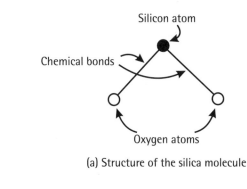

(a) Structure of the silica molecule

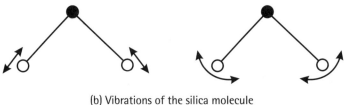

(b) Vibrations of the silica molecule

Figure 3.11 The silica molecule.

longer wavelengths, is absorbed by the molecules of the material, with the effect that the material is heated by the wave. Consequently, its temperature rises very slightly. This type of loss is thus known as absorption loss, and its dependence on optical wavelength is very different from that of Rayleigh scattering, since it only occurs for a small spread of optical frequencies around the various frequencies at which the molecules vibrate. Of course, the glass of the optical fiber is not perfectly pure, so that there are also molecular frequencies of vibration belonging to the impurities, but all are fairly narrow features in themselves. Hence, absorption loss gives rise to a number of peaks in the loss as the wavelength increases, and these may even merge into each other on occasion. So now we have two sources of loss of power from the optical wave traveling in the fiber: Rayleigh scattering and material absorption. These are shown graphically in Figure 3.12, where the variation of the power loss is displayed against optical wavelength. Note the absorption features. This variation is known as the loss spectrum, or attenuation spectrum, because the result of a loss is that the signal power is attenuated. Clearly, we must choose the optical wavelength of the light source for our information transmission with great care. Referring to Figure 3.12: should it be 0.85 μm, 1.3 μm, or 1.55 μm? These are all wavelengths at which the material loss is low, and which lie at the center of what are called low-loss windows, identified in Figure 3.12 as 1, 2, 3.

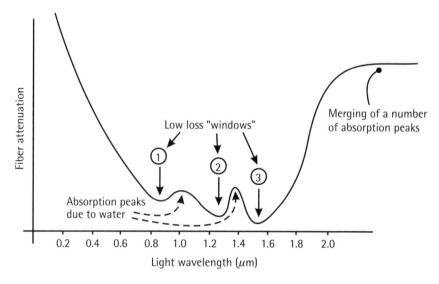

Figure 3.12 Variation of fiber attenuation with optical wavelength.

Before answering this question we must consider yet another effect that results from the light wave's interaction with the fiber material: optical dispersion.

3.4.2 Optical dispersion

We have seen that an optical wave is slowed down when it enters a dense medium, such as glass or water, from air. This was the effect that caused a stick to look bent when inserted end-first into water, and which allowed total internal reflection (TIR) to take place when passing from a more dense to a less dense medium. We know also that the slowing is caused by the wave's encounters with the molecules of the medium: the greater the density of molecules, the more frequent are the encounters, and thus the greater is the slowing-down effect.

We have also seen, in the previous two sections, how the attenuation of the light wave is caused by the wave's encounter with these same molecules, and we have noted how dependent this loss effect is on the wavelength of the light.

It should, therefore, come as no surprise to learn that the slowing effect of the molecules also varies with the wavelength of the light. The simple reason for this is, again, that the strength of the interaction between the wave and molecules depends upon the relationship between the frequency of the wave and the natural frequencies of vibration of the molecule, just as it did in the case of absorptive loss. If the wave frequency is close to a frequency of molecular vibration, the interaction is strong and the wave is slowed by a large amount. Conversely, it is slowed by a much smaller amount when not close to a natural frequency of vibration. Just as we had an optical attenuation spectrum, we now have an optical speed spectrum, and an example of one of these is shown in Figure 3.13. Different light wavelengths travel at different speeds in the fiber.

The consequence of this for optical telecommunications is important, for no light source ever constructed emits only one wavelength. All practical light sources emit light power over a spread of wavelengths known as their spectral width. The spread is broad for a lightbulb and narrow for a laser, but it is never zero. This means, of course, that when the light from the source is injected into the fiber core for guiding, different portions of it (corresponding to the various wavelengths present) will travel at different speeds, and will arrive at the far end at different times. Remember that we chose to use a monomode fiber so that the various mode velocities would not spread the information and hence lead to a reduction in

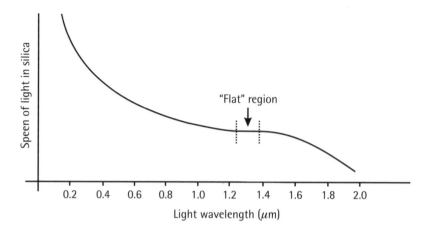

Figure 3.13 Variation of speed of light in an optical fiber with optical wavelength.

bandwidth. Now we find that, even in a monomode fiber, this new wavelength-dependent effect will provide another source of information spreading, albeit less severe than that which occurs in a multimode fiber. This effect again will result in a range of arrival times for different portions of the radiation at the receiver, and thus in the information being dispersed. This dispersion will be greater, the greater the spectral width of the source is. If a light carrier is modulated with a pulse stream, for example, the pulses will tend to broaden and thus merge together as they progress along the fiber. The greater the distance traveled, the greater the merging of the pulses. This will again limit the rate at which they can be transmitted (Figure 3.14); the dispersion effect will, in other words, limit the bandwidth. Since it is due to the properties of the fiber material, the effect is known as "material dispersion." The material dispersion is measured in terms of the maximum delay time per unit of source spectral width, per unit of fiber length. The actual units used are usually picoseconds

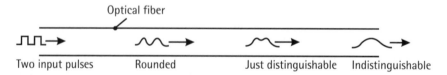

Figure 3.14 Effect of dispersion on optical pulses as they travel along an optical fiber.

($1 \text{ps} = 10^{-12} \text{s}$) for the delay time, nanometers (nm) for the spectral width of the source, and kilometers (km) for the fiber length. This dispersion of the transmitted information can be a serious limitation on the performance of long-distance optical-fiber communications systems unless steps are taken to minimize it.

Looking again at Figure 3.13, it is evident that the speed of light does not change very much when the wavelength of the light lies in the range $1.25 \mu m$ to $1.35 \mu m$. The variation is quite flat in this region. This means that if the light we are using only has a spread of wavelength components within this range, there will be little variation in their times of arrival at the far end of the fiber, and thus little material dispersion. The pulses in a digital system will now suffer very little from spreading distortion. The dispersion (and thus the distortion) will still increase to some extent with the number of wavelength components present, however, and thus with the spectral width of the source. We have, nevertheless, made an attempt to minimize the dispersion.

Now, $1.3 \mu m$ is also a wavelength at which the fiber attenuation is very small: it lies in one of the low-loss windows (see Figure 3.12). So we have both low attenuation and low dispersion around $1.3 \mu m$ (1,300 nm). This is just what is required for high-performance telecommunications. It is for this reason that a whole generation of systems has been engineered to operate at this particular wavelength of $1.3 \mu m$.

However, for the very long distances of several thousands of kilometers (with repeaters, of course) between continents, for example, the low-loss requirement is paramount. So we look toward a wavelength of $1.55 \mu m$ (1,550 nm), where the loss in silica fiber is even lower than at $1.3 \mu m$ (Figure 3.12). Unfortunately, the material dispersion is higher at the $1.55 \mu m$ wavelength, so that the system performance might well now be bandwidth-limited. Fortunately, another dispersion effect actually comes to the rescue here.

The material dispersion effect we have described is a property of the bulk material, in whatever form, and is, therefore, naturally present in the core of a silica optical fiber. However, when the light is propagating in the waveguide structure of an optical fiber, this additional effect comes into play. This effect will now be explained.

As was noted in Section 3.3, when dealing with multimode fibers, the various allowable modes traveled at different speeds down the axis of the fiber as a result of having different reflection angles at the core/cladding boundary (see Figure 3.9b). The same kind of idea is necessary to

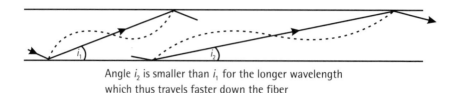

Angle i_2 is smaller than i_1 for the longer wavelength
which thus travels faster down the fiber

Figure 3.15 Waveguide dispersion.

understand this new type of wavelength dispersion. In a monomode fiber, we are, of course, dealing with just one mode, and we know that this mode must satisfy the condition that a whole number of complete wavelengths must fit into the reflection pattern (see Figure 3.9a) for the constructive interference to take place, and thus for the mode to propagate. It is clear from Figure 3.15 that, for this to be so, the reflection angle for the single mode must decrease with wavelength, so that the speed of travel down the axis of the fiber will increase with wavelength. Hence, this is another case of the speed varying with wavelength, and it is now entirely due to the waveguide structure itself: it is thus known as "waveguide dispersion." (Since both material and waveguide dispersion depend upon wavelength, they are sometimes referred to collectively as chromatic dispersion, meaning color dispersion).

Look yet again at Figure 3.13. In the region of the 1.55 μm wavelength, the speed of the light due to material dispersion is decreasing with wavelength, the opposite of the waveguide effect just described. Perhaps it is possible to arrange that the two effects cancel at 1.55 μm? Yes, this is indeed possible. By choosing the waveguide geometry carefully (i.e., core size, and refractive index difference between core and cladding), the two effects can be made to cancel just where we have the third low-loss window. Such fiber is known as dispersion-shifted fiber (DSF), and it is important in the present generation of long-distance systems. Figure 3.16 shows how the total dispersion (material and waveguide) of a monomode fiber varies with wavelength for different core diameters, at a certain value of refractive index difference. At a 6 μm core diameter there is only a small waveguide effect, and the variation differs little from that of the pure material. As the diameter diminishes, the waveguide effect becomes more important and, at 4 μm diameter, it is equal and opposite in sign to the material dispersion at 1.55 μm, so that the total dispersion is zero at this wavelength and for this core diameter: this is, of course, just what is needed. Rarely is there

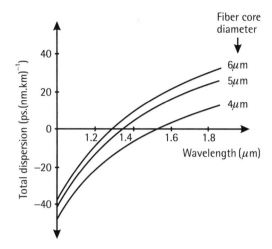

Figure 3.16 Variation of total dispersion (material and waveguide) for dispersion-shifted fiber (DSF).

advantage without disadvantage, however, and there are several cautionary points to note in regard to DSF.

The first is to remember that the dispersion will only be zero at one wavelength: since the source will always contain a range of wavelengths, the overall system dispersion will not be zero, even when operating at the zero-dispersion (ZD) center wavelength (e.g., $1.55\,\mu$m). Hence, there is no such thing as a dispersion-free link, but the dispersion will be minimized (for a given source spectral width) by working at that zero-dispersion wavelength. Second, the refractive index of the core needs to be larger for DSF than for "normal" fiber. This means that more impurities have to be added in order to maintain the monomode condition at this wavelength, and hence there will almost certainly be more optical attenuation. Finally, the core diameter is smaller in DSF, so it is more difficult to launch light into it. All of these features have to be traded off against each other in any system design.

We will now consider some more detailed aspects of this.

3.5 System implications

In order to fix the ideas more firmly in relation to their implications for optical communications systems, let us perform some elementary system design calculations for attenuation (loss) and dispersion. (System design is covered more fully in Chapter 6).

Suppose that we choose an optical wavelength of $1.3\,\mu$m. At this wavelength, which is in the second low-loss window, we may deduce from an attenuation spectrum such as the one shown in Figure 3.12 that the optical power will fall by a factor of about 1.25 for every kilometer (km) traveled in the fiber. Now, a typical fiber system with a moderately sensitive receiver can tolerate a total loss factor of around 100,000, so this means that we can use an unrepeated length of s km of fiber, where 1.25 multiplied by itself s times equals 100,000; that is:

$$1.25^s = 100,000$$

Solving this equation gives $s = 51.6$ km. There will always be losses in the system other than those due only to the fiber, so let us play safe and assume a fiber length of only 45 km. What will be the bandwidth capability of this fiber used as a digital link? Suppose that its dispersion figure is 1.5 ps per nm per km (written 1.5 ps. $(\text{nm.km})^{-1}$). Over 45 km the dispersion will be $45 \times 1.5 = 67.5$ ps. per nm. Suppose now that the chosen source has a spectral width of 2 nm: as a consequence, the total dispersion will be $67.5 \times 2 = 135$ ps. This means that the pulses of light in a digital bit stream cannot be spaced at intervals less than this, if they are not to merge together over this distance of 45 km, and become indistinguishable at the receiver. Now, 135 ps intervals correspond to a bit rate of 7.4 Gb.s^{-1} (i.e., the reciprocal of 135×10^{-12} s). This, then, is the maximum bandwidth capability of the link. Note that its bandwidth-distance product is 7.4 Gb.s$^{-1} \times 45$ km $= 333$ Gb.s^{-1}.km.

It follows that for bit rates above 7.4 Gb.s^{-1} the system is dispersion-limited. For bit rates below that figure it is attenuation-limited. We can thus begin to understand some of the elements of the system designer's problems. The required performance of the link must be matched up to the properties of the fiber, the source, and the capabilities of the receiver. All of these are, of course, variable, within certain limits, depending on how much the designer is allowed to spend. The budget is always one of the most important of the designer's problems.

3.6 Fiber manufacture and cabling

Finally, in this chapter, we shall take a look at the ways in which optical fibers are made, cabled, and installed. The production of the fiber itself depends crucially on the glass composition. The whole of the fiber communications industry depends upon the discovery, made in the mid-1960s, that the absorption of light in glass can be reduced to very low levels by removing the main natural metallic impurities such as iron, manganese, nickel, and cobalt.

The method used to prepare glass of the required purity is to deposit a soot, or powder, from very highly purified gases that are allowed to interact chemically inside a pure quartz tube. The tube is rotated and heated as the deposition occurs, so that the soot is fused into glass as it builds up thickness on the inside of the tube (Figure 3.17). This process (known as chemical vapor deposition [CVD]) also allows for accurate control of the small quantities of added materials (dopants) that are required to form the core/cladding refractive index structure; that is, to make the cladding optical density just slightly less (about 1% less) than that of the core. This is usually done by doping the core with germanium or beryllium, though sometimes by doping the cladding with phosphorous or fluorine, depending on what other properties (e.g., attenuation and dispersion) the fiber needs to have, and at what wavelength it is to operate. When the quartz tube is almost completely filled with its fused deposits, it is removed and cooled to provide what is known as the fiber preform. This preform is effectively a scaled-up version of the final fiber. It is about 1 meter long and 20 mm in diameter. This preform is then fed vertically into a furnace, which melts it, and the thin fiber is carefully drawn from this melt (Figure 3.18). Many tens of kilometers of fiber can be drawn from a single preform. The drawn fiber is immediately given a thin, soft, silicone, primary coating to protect it from attack by the atmosphere (especially moisture) before giving it a much harder, mechanically protective, secondary coating, later. The nature of this second coating varies with the

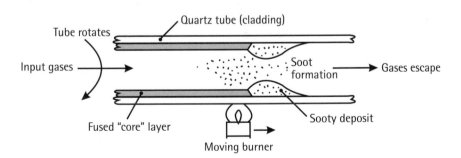

Figure 3.17 Essentials of fiber "preform" fabrication.

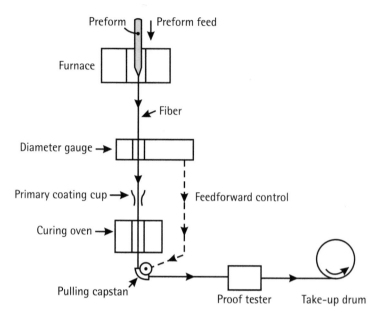

Figure 3.18 Basics of the fiber-pulling rig.

environment in which the fiber is to be used, but it allows the fiber to be handled much like a copper wire.

For use in a telecommunications system, many fibers are usually incorporated into a cable structure (Figure 3.19) for pulling into underground ducts. The pulling process is mechanically demanding, so that the cable has to be reinforced with strength members, as shown in Figure 3.19.

When it is necessary to join together two fibers (or two sets of fibers, in a cable, for example), this is done by placing the ends together and then melting them by means of an arc discharge, thus fusing them together. Such fusion splicing is now a well-practiced art, and is performed with the aid of special fusion-splicing equipment, producing splices with very little loss.

3.7 Summary

In this chapter we looked at the construction of optical fibers and at the principles of light propagation within them. The guiding of light within the optical fiber allows us to use the high-bandwidth capabilities of

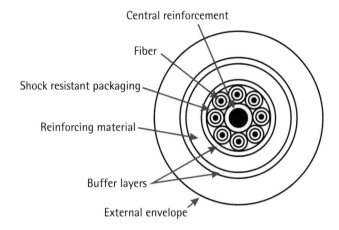

Figure 3.19 A standard optical-fiber cable structure.

the light to convey information, on the light as a carrier, between any two points via a chosen, protected path, over distances of hundreds of kilometers.

We have seen that the light interacts with itself and with the glassy material of which the fiber is made, interactions that limit the performance of the fiber as a telecommunications medium, and that must be well understood in order to provide well-designed communications systems.

In order to gain this understanding, we looked at wave interference, refraction, total internal reflection, and attenuation and dispersion effects in fibers. The principles that have been discussed, together with others that will be introduced as the need arises, will be returned to again and again in the following chapters.

4

Preparing the Light

4.1 Introduction

In Chapter 3 we examined the way in which light travels in optical fibers and noted that, for good telecommunication systems performance, certain properties are desirable for the light. Let us briefly review these:

1. The wavelength of the light must be chosen to correspond to a low-loss window (see Figure 3.12) in order to achieve maximum telecommunications link length for a given light power launched into the fiber.

2. The light power launched into the fiber should be as large as possible (the constraints are discussed in Chapter 7), in order that the fiber length between repeaters should be as large as possible in the face of fiber loss.

3. The light should contain only a small spread of wavelengths in order that a high bandwidth can be maintained over a given length of fiber, in the face of chromatic (i.e., material + waveguide) dispersion (Figure 3.13).

4. The light has to be launched into a monomode fiber core that has a small diameter (5 μm to 10 μm) and at an angle that allows the total-internal-reflection condition to be obeyed in the fiber: the light must, therefore, come from a source of about the same size as the fiber core and, as closely as can be arranged, it should be narrowly confined in its direction so that most of it will obey the total internal reflection condition: in other words the light should be intense (large power from a small area) and highly collimated (i.e., narrowly directed).

5. Since the light has to be capable of carrying large quantities of information very rapidly, there must be available a ready means for impressing high-speed information on to it: we must be able to modulate it at high speeds; for example, by switching it on and off very rapidly, or by passing it through a high-speed shutter.

In summary, then, the light should be intense, well collimated, contain only a small spread of wavelengths, and allow rapid switching (Figure 4.1). These are, in fact, just the properties possessed by light from lasers, so we must look at lasers in some detail.

In addition, it is clear that the sources need to be rugged, reliable, and cheap, because we will need a lot of them and they are to be used in practical telecommunications systems: there should be no parts that are easily breakable or prone to wear, since the devices will need to last for a long time without replacement in what are sometimes quite harsh, remote environments. They may suffer from heat, dirt, and vibration, for example.

Probably the most rugged and reliable of all types of lasers are those called semiconductor lasers. These lasers are also very compact, with an

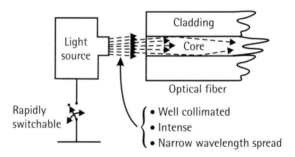

Figure 4.1 Light source requirements.

emitting area of about the same size as a fiber core, and they are readily (and cheaply) manufacturable using well-established semiconductor technology. It is for this reason that semiconductor lasers are almost universally used in present-day optical-fiber communications systems. We need, therefore, to understand how these lasers work. In order to do this it is necessary first to understand the principles by which light is created, and of laser action in general. This will be our first task.

4.2 Light-emission processes

All matter consists of atoms. In some materials atoms are joined with other types of atoms to form molecules: these materials are called compounds (e.g., water, which consists of molecules that each contain one atom of oxygen and two of hydrogen). In other materials, only one type of atom exists: these materials are the elements, such as hydrogen, oxygen, silicon, and copper. Atoms are extremely small: the largest are only about 1 nanometer (1 nm; 10^{-9} m) in diameter. An atom can most simply be pictured as a structure that contains a central, positively charged nucleus with a number of negatively charged electrons circling around it, like planets around a sun (Figure 4.2a). The total positive charge on the nucleus results from the number of protons it contains, each proton having exactly the same amount of positive charge as an electron has negative charge. The number of circling electrons is equal to the number of protons in the nucleus, so that the atom as a whole is electrically neutral. It is this number of protons (or electrons) that determines the chemical properties of the atom, and that thus defines the element to which it corresponds: this number is known as the atomic number. Each element thus possesses a characteristic atomic number; for example, hydrogen (1), silicon (14), iron (26), copper (29), silver (47), and so on. The electrons encircle the nucleus at various distances, and each atom has its own special set of distances, fixed by its atomic number. The distance at which any given electron circles the nucleus is, not unreasonably, called the radius of its orbit. The energy possessed by any given electron in its orbit is determined largely by the radius of its orbit: the larger the orbit, the greater is its energy. (The actual values of the radii and the electron energies can be worked out in detail using the principles of quantum theory, but these need not concern us here.)

Now, a given electron can move from one orbit to another under certain circumstances, which we will discuss shortly. Clearly, when it does so

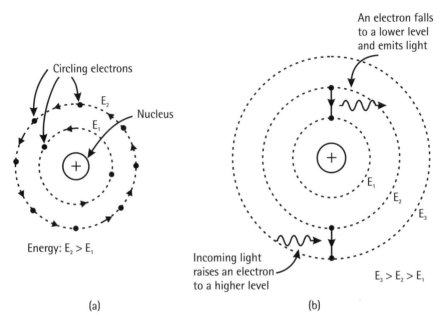

Figure 4.2   Basics of atomic structure and light emission/absorption processes.

move, its energy must change (Figure 4.2b). If it moves from a higher to a lower orbit, its energy decreases, and since energy can neither be created nor destroyed, the energy it has lost must appear in another form. Quite often this energy appears as light. This is one way of creating light. Conversely, when the electron moves from a lower to higher orbit, its energy increases, and this energy must be provided by an external source. Again, quite often, this energy can be provided by light coming in from outside the atom (see Figure 4.2b).

Let us consider in more detail the case where the electron falls from a higher to a lower level. If it is to fall to a lower level, there must be room for it at this lower level. If left to itself for a time, any atom will adopt its lowest possible state of total energy, as will any physical system. A swing, for example, will come to equilibrium with the seat at the lowest point and the ropes vertical: energy has to be put in (i.e., someone has to push) to move it up to a higher energy state. If the swing seat is pushed to a higher level and then released, the seat will swing back to the lowest point, and the energy it had as a result of the push is released in the form of the energy of motion the seat has as it passes through the lowest point (see Figure 4.3). Similarly,

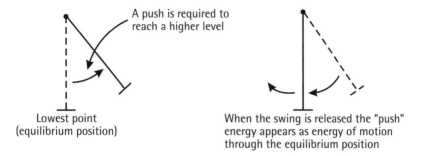

A push is required to reach a higher level

Lowest point
(equilibrium position)

When the swing is released the "push" energy appears as energy of motion through the equilibrium position

Figure 4.3 Swing dynamics.

with the atom, the lowest energy state (the equilibrium state) for the atom as a whole consists (broadly speaking) of an arrangement of electrons which fills up the lowest states first and builds upwards (Figure 4.4). Now, the higher the energy level, the more electron states there are available, since the circumferences of the higher orbits are larger, and there is thus more room for electrons. In fact, if the orbits are numbered in order, starting with the smallest, as 1,2,3, ... and so on, then for the nth orbit the number of available states is $2n^2$. Hence the smallest orbit has 2 states (2×1^2), the second smallest has 8 (2×2^2), then 18 (2×3^2), 32 (2×4^2), and so on (see Figure 4.4a). These states are filled from the lowest up, but only for an isolated atom in equilibrium; that is, one that is not receiving any energy "pushes" from outside. However, atoms are not isolated from each other in a material. In a gas they are constantly moving, largely independently but occasionally colliding; in a liquid they are continuously sliding over each other; and in a solid they are fixed, but quite close together and constantly vibrating against each other. All of these movements are a consequence of the temperature. The greater the temperature, the greater the energy of movement of the atoms, in any state of matter. All of these movements are random. This constant interaction of the atoms in a material means that they are continually being disturbed, by pushes, from their overall states of lowest energy. These are the states where the available positions are filled with electrons from the bottom up. Now, the collision energy, which depends directly on the temperature, causes some electrons to be knocked upward from their lowest possible levels into the higher, unoccupied levels. This, in turn, creates vacancies at the lower levels to which they can return. Hence there is a constant agitation of upward and downward movements with, at any one snapshot in time, a distribution

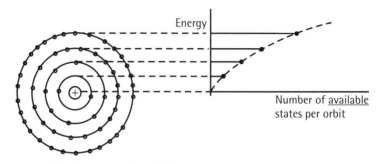

(a) The number of available states increases with energy level

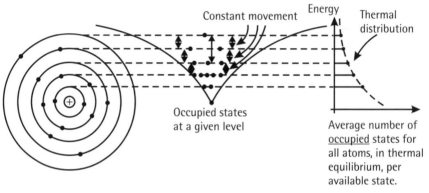

(b) Energy distribution for an interacting atom in a gas in thermal equilibrium

Figure 4.4 Atomic electron distributions.

of electrons among the various possible energy levels, which is constant for a given temperature. This distribution will have steadily diminishing numbers of electrons per available state at the higher energy levels, because the higher the level, the less likely it is that any given electron can be "agitated" up to it by the random movements that result from the temperature. This snapshot equilibrium is clearly a dynamic equilibrium since the individual electrons are constantly moving up and down, but the overall numbers in each state remain constant in time: this is called thermal equilibrium (Figure 4.4b). The important point for our present purposes is that the downward movements of electrons can, and often will, emit light (Figure 4.5). The upward movements, which are the result of the energy input that is maintaining the temperature, are thus essentially being

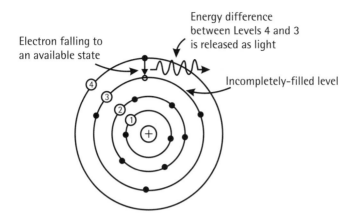

Electron falling to
an available state

Energy difference
between Levels 4 and 3
is released as light

Incompletely-filled level

Figure 4.5 Atomic light emission.

converted into light. This is why a poker glows red when taken from a hot fire, and why the sun gives us heat and light (in this latter case, the temperature is maintained by nuclear reactions in the sun's core). Also, we learned in Chapter 1 that an electric current is just a flow of electrons, and when a current is passed through, say, tungsten (74) the flowing electrons agitate the tungsten atoms and raise the temperature of the material, causing it to emit light and heat. This is the principle of the electric lightbulb: it gives us light, and we know very well that it also becomes hot to the touch.

Since many different energy levels are involved, the radiated energy is emitted over a broad range of wavelengths, some at optical wavelengths, some at infrared and ultraviolet wavelengths, and beyond. In fact, all wavelengths are represented to some degree in the radiation from a body of material at a given temperature (Figure 4.6). There is a definite relationship between these energy levels and the wavelength of the emitted radiation, which we shall now explore.

What, exactly, is it that happens when an electron falls from a higher- to a lower-energy orbit? When this occurs for a single electron in a given atom, a particle of light is emitted (Figure 4.7). This particle is called a photon. But isn't light a wave? Yes it is, but it also consists of particles. This so-called wave/particle duality runs through the whole of physics. Sometimes, for the processes of wave propagation and interference, for example, light is best regarded as a wave; at other times, as here, in its interaction with matter, it is best regarded as a stream of particles: a stream of photons.

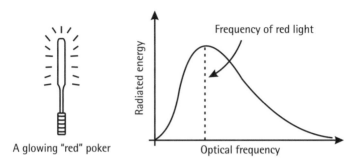

A glowing "red" poker

Figure 4.6 Radiation from a hot body.

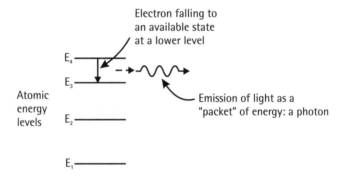

Figure 4.7 Photon emission.

At the simplest level, the wave can be viewed as a guide for the photons, controlling their distribution and direction (Figure 4.8a).

We can view the electron falling from a higher-energy level to a lower one as releasing the excess energy in the form of a particle of light (a photon) that has that excess energy. The larger the energy released, the higher the frequency of the light wave associated with that photon, so that large energy reductions will give ultraviolet light, medium reductions visible light, and small ones infrared light (Figure 4.8b). This can best be pictured as energy of wave agitation. The higher the frequency, the more energetic is the wave's agitation, and vice versa (Figure 4.8c). In fact, the energy of the photon is directly proportional to the wave frequency: twice the frequency means twice the energy for the photon, for example. (The exact relationship is $E = h \times f$, where E is the energy, f is the frequency, and h is a fundamental constant known as Planck's constant, with a value of 6.625×10^{-34} joules.seconds.)

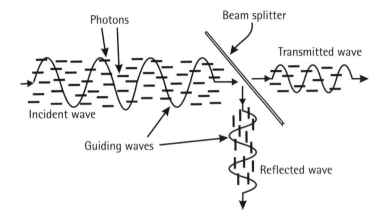

(a) Photons and their guiding waves

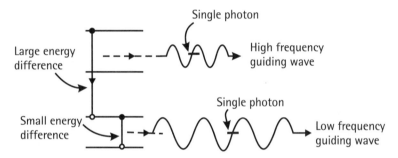

(b) Relationship between atomic electron energy changes and the emitted light

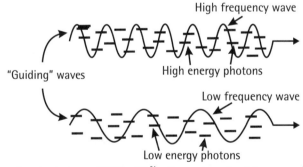

Energy of each photon = $6.625 \times 10^{-34} \times$ frequency of guiding wave ($E = h \times f$)

(c) Relationship between photon energy and optical frequency

Figure 4.8 Wave/particle duality.

Consider now the reverse process of an electron moving upward from a lower- to a higher-energy orbit. In this case, the required extra energy must be provided from somewhere, and we have seen that one way of providing this is to bombard the atom with electrons, by passing an electric current into the material. Another is to bombard it with other atoms or molecules, and one way of achieving this is to heat the material. A material is heated by putting it into contact with another body at a higher temperature, as, for example, the poker in the fire, or the cake in the hot air of an oven. A higher temperature merely means that the molecules are moving more energetically; that is, they have higher energy (Figure 4.9). So they bombard the atoms or molecules of the cooler body and increase their energies; in other words, they heat up the cooler body. In so doing, some of the electrons are moved into higher-energy orbits, and can emit light as they move down again. This is why a hot poker glows red, and an even hotter one glows white (strictly, it is closer to blue). These colors correspond to the dominant frequencies for the two different temperatures (see Figure 4.9).

Yet another way of providing the energy is to bombard the atoms with photons; that is, with light. A photon with a particular energy will be able to lift an electron from a lower-energy orbit to a higher orbit but, and this is an important point, this will only happen if the atomic structure has two orbits which differ in energy by an amount that is exactly equal to the energy of the incoming photon. If this is not the case, the photon will scarcely interact with the atom at all (see Figure 4.10a). Of course, it is not necessary for the energy jump, in either direction, to take place between two immediately adjacent, or even close, levels. If, for example, the incoming photon, of a certain wavelength, has sufficient energy to stimulate a

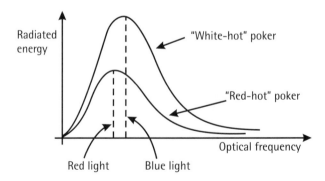

Figure 4.9 A body radiates more energy at a higher temperature.

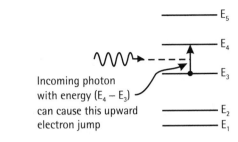

(a) Incoming photon causes upward transition

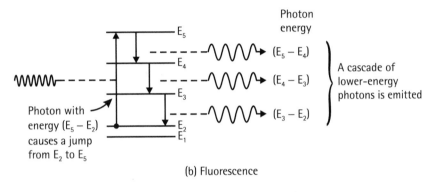

(b) Fluorescence

Figure 4.10 Upwards transitions caused by incoming photons.

jump from level 2 to level 5, say (see Figure 4.10b), this jump can also occur. The electron can then return to level 2 via levels 4 and 3 (see, again, Figure 4.10b), and emit photons of smaller energy (larger wavelength, smaller frequency) as it does so. This can lead to the phenomenon of fluorescence, where high-energy photons of, for example, ultraviolet light, can cause a material to glow at longer, visible wavelengths.

Having now understood the way in which photons and atomic energy levels interact, we can appreciate how materials can emit light when some kind of energy is pumped into them, and how the materials also can absorb light. We are now in a position to understand how a laser works.

4.3 Laser action

Suppose that we have a particular element in the form of a volume of gas at a particular temperature. Suppose also that each atom of this element has, for its circling electrons, a set of energy levels (corresponding to different sizes of orbit) as shown in Figure 4.11. We know, from the discussion in the previous section, that if a photon comes in from the outside, with energy equal to the difference between any two of these levels, and strikes the atom, then an electron will be raised from the lower of the two levels to the upper one (provided that there is room there, so that the upper one must lie in the upper reaches of the equilibrium distribution). Hence a photon with energy $(E_3 - E_1)$ will cause an electron to rise from level 1 to level 3, and a photon with energy $(E_2 - E_1)$ will cause one to rise from E_1 to E_2 (Figure 4.11). Suppose that the first of these two possibilities actually occurs, and that we allow a large number of photons with energy $(E_3 - E_1)$ to strike the gas. Large numbers of atoms will have electrons moving from

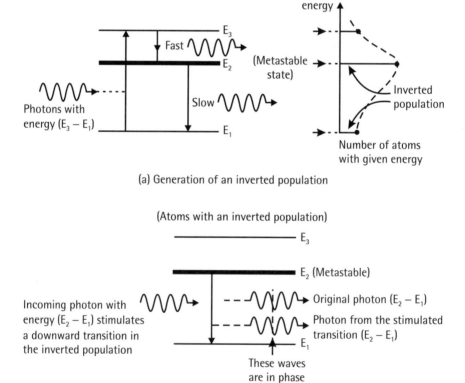

(a) Generation of an inverted population

(Atoms with an inverted population)

(b) Stimulated emission

Figure 4.11   Inverted population and stimulated emission.

level 1 to level 3. The electrons will not be comfortable at this high level (3), since there are states at the lower levels into which they can fall, especially level 2, where there is plenty of room. So they fall quickly into level 2. Level 2 is deliberately chosen to be a rather special state: what is known as a metastable state. This means that the electrons are relatively happy to remain there for a while even though there is still room in level 1. The reasons for this lie deep in quantum theory, and are concerned with the ease with which any transition between levels can be made. The transition 2→1 is difficult, so the electrons linger in level 2, but only for, perhaps, a few hundred microseconds. This is quite long on atomic timescales, and it is certainly long enough for the higher level 2 to become much more heavily populated with electrons than the lower level 1 (remember that level 3 is being filled with electrons continuously, from level 1, by the incoming light). This is now a reversal of the state of thermal equilibrium, where higher energy levels always contain fewer electrons per available state than lower ones (see, again, Figure 4.4b): it is thus known as an "inverted population" (Figure 4.11a).

The inverted population is a crucial feature of laser action. To see why, consider what happens when a photon with energy $(E_2 - E_1)$ enters this system and interacts with one of the inverted-population atoms. This photon can now actually stimulate the electron to fall from state 2 to state 1. The electron in state 2 has a higher energy than it has any right to have, and it is looking for any excuse to fall to state 1; it is just that those tiresome quantum rules are deterring it from doing so. The incoming photon, entering the system with just the right energy and having a healthy disregard for localized rules governing the behavior of electrons, provides a convenient nudge, and the electron is stimulated to make the downward jump. This jump occurs almost immediately after the incoming photon strikes the atom. The downward transition is therefore now very fast, taking place in, typically, a few nanoseconds (ns). Because the downward jump means a loss of energy, this energy appears as a photon. So the incoming photon has given rise to a second photon of the same energy, wavelength, and phase. The first photon has stimulated the emission of a second: one photon has become two. Further, the guiding waves of the two photons are locked together, just as the person pushing the swing must lock his or her efforts to the cycle of the swing's motion if a good swinging action is to be obtained. This means that the guiding waves will be of the same phase. The complete process we have just described is called stimulated emission (see Figure 4.11b).

Moreover, when there is a large number of atoms within the inverted population (as there will be in any sizeable volume of material), a flux of about as many photons, each with energy $(E_2 - E_1)$, entering into the system will, by means of this process, result in an even larger flux of photons, each with that same energy. The incoming light has thus been amplified. We have light amplification by stimulated emission of radiation, or laser, action.

Remember that the light with photons at energy $(E_2 - E_1)$ has been amplified by means of the inverted population, itself produced by incoming light with photons at energy $(E_3 - E_1)$. This latter light is known as pump light. The system is pumped with light at $(E_3 - E_1)$, one wavelength, to provide laser amplification for light at $(E_2 - E_1)$, another wavelength.

This amplifying action is very useful in itself. We can use it to devise an optical amplifier; that is, a device that will take in light (with a wavelength that corresponds to the population-inversion energy difference) at its input and produce an increased power level, at the same wavelength, at its output. Such optical amplifiers are very useful in long-distance optical communication systems, and we shall be looking at them in more detail in Chapters 6 and 7. Our present concern is with light sources, however, so let us now see how a laser light source can be constructed, using the above process.

In Figure 4.12 we can see the basic structure for a laser. The lasing material (the volume of gas we have been considering, say) is placed within a container that consists of a glass cylinder with its two ends enclosed by plane, parallel mirrors. The material within this enclosure is now exposed, from the side, to a flux of pump light whose photons have energy corresponding to the transition between states 3 and 1 in the gas; that is, $(E_3 - E_1)$. Many of the atoms in the gas are now pumped up to state 3, fast

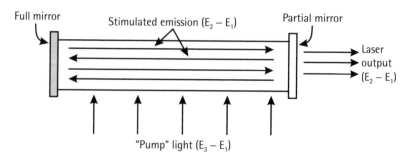

Figure 4.12 Basic laser construction.

decay then takes place to level 2, and an inverted population is produced, all as was previously described.

Suppose now that one of the electrons in the metastable (long-lifetime) state 2 decays to level 1 (remember that a long lifetime in the atom is still only about 300 μs). Since it is falling to a lower state, a photon of energy $(E_2 - E_1)$ is emitted. This photon then can impact upon an adjacent excited (i.e., member of the inverted population) atom to cause it to fall from E_2 to E_1 by the process of stimulated emission. One photon has become two, as before: fast amplification has again occurred. The only difference between this case and the one described earlier is that the photon at $(E_2 - E_1)$ has, in this case, been produced naturally within the system, rather than having entered from the outside. Clearly, this process can now continue: two photons will become four; four will become eight, and so on. Furthermore, the process is enhanced by the mirrors at the ends of the container, since these will reflect back any photons that reach the ends of the cylinder, and these can then continue with the process of stimulating the emission.

So we have, as long as we continue to pump this system, light with a narrow range of frequencies (corresponding to the energy $(E_2 - E_1)$) bouncing between the two mirrors in our cylindrical container. If the cylinder has a small diameter and it is quite long, there is then a large light power (i.e., many photons per second) all traveling along one line, in both directions, all with pretty much the same frequency, and all contained within a small cross-sectional area. Moreover, since the stimulated emission process produces photons whose guiding optical waves are all in step with each other, the light can be described by just one guiding wave—it all has the same phase (Figure 4.13). This is going to be very useful light:

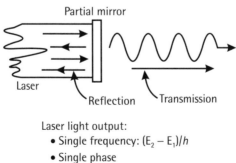

Laser light output:
- Single frequency: $(E_2 - E_1)/h$
- Single phase
- Highly collimated
- Intense

Figure 4.13 Characteristics of the laser light output.

intense (because large power over a small area); of narrow frequency/ wavelength spread (because it corresponds just to the energy transition $(E_2 - E_1)$); highly collimated (because it is bouncing between two parallel mirrors); with almost a single phase (because successive stimulated emissions are all in step); and rapidly responsive (because the stimulating photons cause the downward transition to occur much more rapidly than it otherwise would). All we have to do now is make one of the two end mirrors partially transmitting (recall, from Chapter 3, that light striking glass from air, for example, will lead to both a reflected and a refracted ray), and some of this light will come out of the container. Then we can use it. We have constructed a useful laser. The basic theory of laser action is presented in more formal terms in Appendix D, for readers who may be interested.

The properties of this laser light are, of course, useful for a large range of applications, from reading bar codes in supermarkets and CD-ROMs, through laser surgery and industrial welding, to "Star Wars" anti-missile defenses. But it is also very useful for optical-fiber telecommunications: it has just the required properties described at the beginning of this chapter, provided that the laser source can produce light at the wavelengths required for low attenuation and low dispersion, and provided that it can be small, rugged, and cheap. All of these requirements are satisfied by a particular type of laser known as a semiconductor laser.

4. 4 The semiconductor laser

Solids, liquids, and gases represent different states of matter. Almost any physical substance can exist in any of these states: for example, water can exist as ice, tap water, or steam. Which state it is in at any given time depends primarily upon its temperature, for temperature, as we have already noted, is a measure of the energy of the atoms or molecules that compose the material. The greater is this energy, the greater is the ability of the atoms and molecules to break free from their mutual attraction, which tends to try to bind them securely together. Thus, when the temperature is low enough, the binding forces predominate, and the material hangs firmly together as a solid; when it is high enough, the energies of motion of the individual atoms and molecules predominate, and they fly about freely, as in a gas; between these two states there is an intermediate condition where they are not totally free but not quite firmly held, so that they slither and slide across and around each other, and we then have a liquid (Figure 4.14).

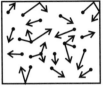

Gas	Solid	Liquid
(Molecules move independently, occasionally colliding)	(Molecules in fixed relative positions)	(Molecules are semi-independent: they slide over each other)

Figure 4.14 States of matter.

In a solid, the molecules are firmly fixed with respect to each other and are thus densely packed. The result of this is that the sharp energy levels, of the electrons in their atomic orbits in the gas that was discussed in the previous section, now overlap and interfere to form bands of energy. The atomic electrons now occupy these bands rather than the well-defined orbits of the atoms in a gas (Figure 4.15). When the uppermost of these bands is only partly filled with electrons, there are plenty of immediately-adjacent energy spaces for the electrons to move into with almost no addition of energy, unlike the case in the gas where a definite injection of energy was required before an electron could rise from one state (E_1) to another (E_2). Thus, in this case of partial band filling, the electrons can move easily around the material under the action of quite small forces, and the material is an electrical conductor, a metal such as copper or silver, for example (Figure 4.16a). If one of the bands is completely full with electrons, and the band above it is completely empty, then the reverse is the case: the electrons cannot move at all, since they need a lot of energy to move up to vacant energy states in the band above (Figure 4.16b); such a material is an electrical insulator: glass or porcelain, for example. Finally, and most important from the point of view of the present discussion, there is an intermediate state where one of the bands is completely full and the band above it completely empty (as before), but now the gap between these bands is quite small (Figure 4.16c). It is so small, in fact, that only a small amount of energy is needed to raise some electrons from the lower to the upper state, an amount of energy that can be provided by a small rise in temperature, for example. Once in the upper, empty, band the electrons can again move quite freely, for there are many levels for them to occupy, and so the material can then conduct electricity. However, since only a relatively small

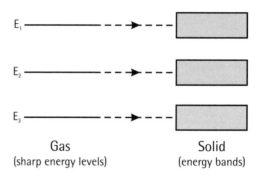

Figure 4.15 Energy bands in solids.

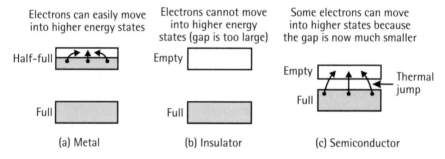

Figure 4.16 Energy-band structures for metals, insulators, and "intrinsic" semiconductors.

number of electrons can rise to the upper band, the electrical conduction is still quite small, much smaller than in the case of a metal (but much larger than for an insulator), and the material is thus known as a semiconductor.

One further step is required, and then we are in a position to understand not only semiconductor lasers but also the basis for the whole of modern electronics, including TV, computers, and mobile telephones. This step concerns the controlled addition of certain materials, known as dopants, to natural semiconductors of the kind described above. The action of these dopants is to modify the band structure, to our advantage. There are two types of dopant: the first provides some easy electrons to the upper band by, effectively, injecting populated energy levels just below the upper band (Figure 4.17a). This increases the number of electrons in the upper level and thus increases the conduction, bringing it even closer to the metallic state. Since electrons are provided, and these have negative

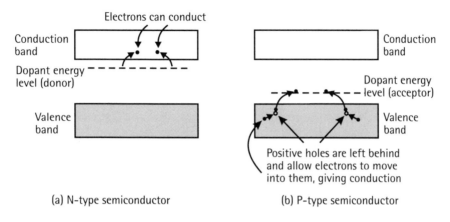

(a) N-type semiconductor   (b) P-type semiconductor

Figure 4.17 "Extrinsic" semiconductors.

charge, this is called an n-type semiconductor. The other type of dopant, conversely, provides some vacant energy levels just above the lower band (Figure 4.17b). This has the effect of allowing some electrons from the lower band to rise easily into those levels and hence create vacancies, in the lower band, known as holes. These holes increase the conduction of the material because they allow electrons in the lower band to move more freely, since there are now spaces into which they can move. These spaces are, effectively, absences of negative charges and are thus equivalent to positive charges: this type of material is called a p-type semiconductor.

One of the prime advantages of doped semiconductors is that we have the conduction under direct control, via the choice of doping level. Natural semiconductors, without any doping, are called intrinsic semiconductors, in order to distinguish them clearly from the doped variety, which are called extrinsic semiconductors.

Very interesting things happen when p-type and n-type materials are brought together, in what is known as a p–n junction. (Modern electronics depends upon these things, but we cannot be concerned with many of them here since our task is specifically to understand semiconductor lasers).

A light-emitting diode (LED) is a p–n junction where the electrons from an n-type material fall into the holes in a p-type material. The electrons are provided continuously by passing an electric current (stream of electrons) into the p–n junction (Figure 4.18a). Clearly when the electrons fall from the higher level to the lower level they lose energy and, provided

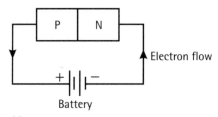

(a) Electrons are passed into a P–N junction

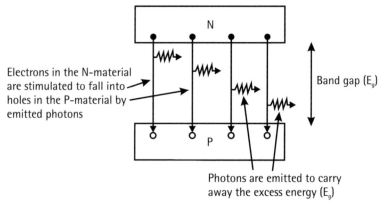

(b) Light emission from an electrically-powered P–N junction

Figure 4.18 The P–N junction.

we choose our material correctly, this energy will appear as photons; that is, as a light emission. So what we have in a p–n junction is a physically constructed inverted population. As soon as the two types of material are brought together, there is a set of populated states (electrons in the upper band of the n-material) of higher energy than vacant states into which they can fall (holes in the p-material). The difference in energy between the two sets of states is called the band-gap energy and this, of course, will determine the wavelength of the light output: often this is in the red or infrared regions for an LED.

Stimulated emission also occurs because the emitted photons will assist in this process of combining the electrons and the holes by stimulating the electrons to fall quickly. All we need to do now is coat the ends of a slab of material with partially reflecting layers (or even just polish them), to provide the mirrors for the back-reflection process described in

Section 4.3, and we have a semiconductor laser (Figure 4.19). Of course, the electrons in the upper layer of the n-type material must be replenished continuously from the electrical power source, and this is now the electron pump for the laser (e.g., a battery).

The semiconductor laser is ideal for optical-fiber communications. First, it is small in size so that the light-emitting area can match the size of the optical-fiber core, and thus launch most of the light into the fiber. Second, it is rugged: a small, solid chip with no moving parts or empty spaces. Third, it is easily switched on just by passing a current into it from a low-voltage source (1–3v). Probably most important of all, however, is the control we have over its optical properties via the choice of materials, the dopant levels, and the geometry. This control has led to a range of valuable devices at chosen wavelengths, with designer power levels, wavelength spreads, and physical sizes.

The structure of a modern semiconductor laser designed for optical communications use is illustrated in Figure 4.20. It looks quite complicated, but each feature has an easily understandable purpose. The lasing material, gallium arsenide in the illustration, is an alloy of the elements gallium and arsenic, and has the advantage that the material can readily be made n-type or p-type simply by varying the relative amounts of each element. Also, it has a band gap whose corresponding wavelength lies in the first low-loss fiber window, at $0.86\,\mu$m. Other elements such as indium and phosphorous can be added to the mix in various proportions to increase the range of optical frequencies available, in order to meet the requirements of modern optical communications systems. These materials operate at longer wavelengths, in the other two windows ($1.3\,\mu$m and $1.55\,\mu$m). All of these materials allow very fast and efficient combinations of the electrons and the holes from the p-type and n-type materials, so that a large

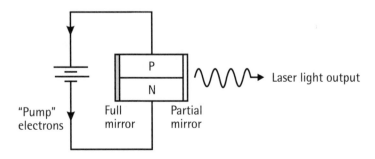

Figure 4.19 Basic semiconductor laser construction.

power output can be obtained from a relatively small current input, and very little input power is wasted as heat.

The lasing layer of gallium arsenide in the illustration (Figure 4.20) is a thin one (its thickness is about 0.2 μm) sandwiched between the pumping layers of gallium aluminium arsenide. This effectively makes the central layer a waveguide, since it has higher refractive index than the surrounding layers. This serves to confine the optical beam and reduce losses to the sides of the structure.

Next to the pumping layers of gallium aluminium arsenide are similar, but more heavily doped, layers of this material. These, as a consequence of their heavy doping, are highly conducting, and act as a smooth, crystal-compatible transition between the lasing regions and the metal electrodes by which the pumping current is injected. The silicon dioxide acts as an insulating structure to channel the injected current effectively to the lasing region so that few electrons are lost sideways, and heating of the device is thus reduced. Even so, the heat sometimes has to be removed by a heat sink (a block of material that carries the heat away) to protect the device from overheating. The device overall is quite small. The front face is about 100 μm × 100 μm (i.e., 0.1 mm × 0.1 mm) and the laser-emitting area is of the same order of size as the fiber core (about 5 μm in diameter). The emitting face is usually polished, to provide a good reflecting surface for the laser action, and then glued (with carefully chosen glue) firmly to the (also-polished) fiber end, to form a good optical join. Thus the intense laser light emerges from the small emitting area in a parallel beam within

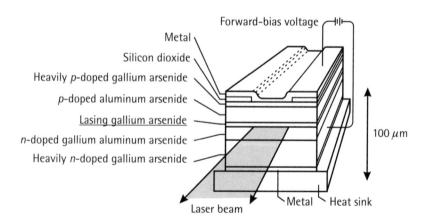

Figure 4.20 Design for a GaAs heterojunction semiconductor laser.

the limits of the core area; this is, of course, just what is required for efficient launching of the light into the fiber core.

We have already noted that the semiconductor laser possesses the important advantage that semiconductor fabrication technology is very well established for the manufacture of electronic chips and that, therefore, these lasers can be produced reliably and in large numbers, relatively cheaply. A corollary of this is that the semiconductor laser can also actually be integrated on to a chip which comprises other associated functions such as the current (pump) feed, automatic monitoring and control of the laser output, temperature monitoring, modulation circuitry, et cetera. In other words, this form of laser is entirely compatible with modern electronics.

The semiconductor laser, which is sometimes known as a semiconductor laser diode (SLD), since it is a two-electrode (diode) device, thus comprises a very suitable light source, feeding into the optical fiber which will then convey the light with low attenuation and low dispersion over large distances.

Two other things are needed before we have an effective basic telecommunications link. The first is a means for impressing large quantities of information on the light carrier before launching into the fiber; the second is a means for detecting the light at the other end of the fiber together with a means for extracting the impressed information. We will look closely now at how we can impress information on the light—the process of optical modulation—since this is part of the light preparation process. Light detection is the subject of our next chapter.

4.5 Optical modulation

The process of impressing information on a light carrier is known as optical modulation, as already noted. It involves varying one or more of the features that define the optical wave, such as its power level, frequency, or phase. In the case where the source of the light is the semiconductor laser, by far the easiest modulation is that of the power. The amount of light power that emerges from the device depends on the electric current that is pumped into it. The greater the current, the more electrons there are in the upper band of the n-type material, and the greater is the number of photons that is produced on their recombination with holes.

Unfortunately, there is not a direct, proportional relationship between the current and the output light power: doubling the current does not necessarily double the power, for example. Consequently, the output power does not, in general, faithfully reproduce the current signal (Figure 4.21).

The difference between analog and digital signals was explained in Section 1.6. The analog signal accurately corresponds with the original information: an electron current will rise and fall in exact sympathy with the changes in pressure that comprise the sound wave of a speech signal, for example. Clearly, the output light power will not reproduce this signal if such an electric current is used to pump a semiconductor laser. However, we have also seen how an analog signal can be digitized; that is, turned into a coded stream of pulses (Figure 1.10). This form of signal is ideally suited for the semiconductor laser, for one pulse of current will always produce one pulse of light, and there is thus a one-to-one correspondence between the input current pulse stream and the output optical pulse stream (Figure 4.22).

Effectively, we are using the electric current, which contains the information to be transmitted, to switch the laser on and off very rapidly. How rapidly can this be done? This depends, fundamentally, on how rapidly electrons and holes can be made to recombine.

Now, we have already noted that one of the advantages of using a laser as a light source is that it relies on the process of stimulated emission, which is a very fast process. In this process, very large numbers of photons, which are bouncing between the different layers of material and the end reflectors, are forcing the electrons in the upper state of the n-material

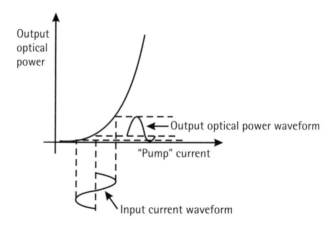

Figure 4.21 Non-linearity of the semiconductor laser characteristic.

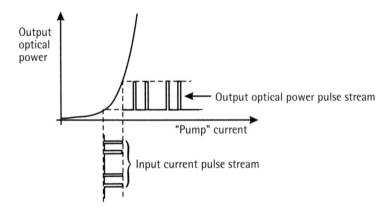

Figure 4.22 Faithful reproduction of a pulse stream with a semiconductor laser.

down to join with holes in the lower state of the p-material. They are thus recombining much more quickly than they would have done if left alone. The laser action is, then, by its very nature, a fast one, and we can expect a rapid optical response if a current pulse is suddenly injected into the laser.

However, the laser materials must be chosen carefully. First, they must be chosen so that the energy gap (i.e., the band gap) between upper and lower levels corresponds to the optical wavelength required (e.g., 1.3 μm or 1.55 μm). Secondly, they must be chosen so that the recombination of electrons and holes is fast, even by lasing standards: this is why gallium arsenide and various alloys are used. Gallium arsenide is what is known as a direct band-gap semiconductor, meaning, broadly speaking, that the recombination is a simple interaction between the electrons and the holes, not involving any other parts of the material. This means that it can be very fast, less than 1 ns in duration, so that the device can be switched on and off at rates up to about 5 Gb.s^{-1}. Finally, the materials must be chosen so that they interface readily with each other in a structure such as that shown in Figure 4.20. This implies that the arrangements of the atoms in each of the materials should be much the same; that is, should have the same kinds of crystal structure. This allows a stable, strain-free overall structure that provides clean, precise, optical features. Such a structure is called a "heterostructure."

Clearly, then, we are now able to arrange that a semiconductor laser, with all its other desirable properties, can also switch on and off very rapidly in response to a fast digital stream of current pulses, in order to produce a correspondingly fast (about 5 Gb.s^{-1}) stream of optical pulses for efficient launch into the core of an optical fiber. Digital signals, as we have learned, have important advantages of their own: it is much easier for a detector system to recognize just the presence or absence of a pulse at a particular time than to determine the exact level, within a narrow band, of an analog signal at that time.

Almost all optical communications systems, and certainly all trunk systems, are digital. This means that the semiconductor laser is an ideal source from this point of view also. It is, therefore, almost universally used in these systems at the present time.

Sometimes, nevertheless, analog modulation is desirable. There are some very effective ways by which laser light, including that from semiconductor lasers, can be analog modulated. To understand these methods we need to review some more optical ideas and these, plus the modulation methods, will be covered in Chapters 6 and 7.

4.6 Summary

In this chapter we have emphasized the requirements for the kind of light that is needed for optical-fiber communications. These requirements are best met by laser light. We have studied light-emission processes, leading on to an understanding of basic laser action and construction.

The type of laser best suited to practical optical-fiber communications is the semiconductor laser, and in this chapter we have discussed, broadly, what a semiconductor is and how its properties can be harnessed to design a semiconductor laser. This turns out to be an almost ideal source for digital optical-fiber communications systems, and it is almost universally used in present-day systems.

5

Seeing the Light

5.1 Introduction

Having impressed the information on the laser light and launched it into the core of the fiber, the light propagates through the fiber, suffering attenuation and dispersion as it does so. We have looked at the effects of attenuation and dispersion, and we know that the former reduces the level of the signal while the latter distorts it, so that the distance over which the information can be transmitted reliably is limited by whichever one of the two first renders the signal unacceptable at the receiver.

What is or is not acceptable at the receiver also depends upon the receiver itself, so it is important for us to look at the process of light reception. After doing so, we will be able to see where the reception processes fit into the overall communications system design.

The information arrives at the receiver in the form of a modulated light wave, usually a coded stream of optical pulses. The pulses will, by this time, be at a fairly low level, as a result of the fiber attenuation, and they also will have broadened out somewhat, as a result of the fiber dispersion (see Figure 3.14). The receiver's first job is to reproduce these pulses accurately as an electrical current signal; that is, to convert them into a corresponding stream of electrical pulses. When the pulse stream is in electrical form, we can then use all the facilities of modern electronics to extract the required information, and do with it whatever we wish. We will return to these matters later, but for the moment our task is to look at the way in which an optical signal can be converted into an electrical one. This is what an optical detector, or photodetector, does.

It is fairly easy to state the basic requirements that an optical communications link demands of such an optical detector. First, it should be quite sensitive; in other words, it should be able to deliver usable amounts of electric current for quite small optical power inputs at the preferred optical wavelengths of 1.3 μm or 1.55 μm. Second, it should be fast: it should be able to respond to the fast optical pulse stream sufficiently rapidly to be able to reproduce it accurately in electrical form. Finally, it should be physically small, rugged, cheap, and consume only a small amount of electrical power. These latter requirements, as for the light source in the last chapter, are necessary for practical, reliable communications links that must also be economic, since large numbers of these photodetectors will be required for installation in sometimes quite hostile environments (in repeaters, for example). Hence, for the same reasons as before, we look again toward semiconductors to provide suitable photodetection devices.

5.2 The photodiode

A beam of light is a flowing stream of photons. Each photon has a definite energy, so the optical power flowing into a photodetector is equal to the rate at which photons are arriving at its surface multiplied by the energy per photon. The power is the rate of arrival of energy.

In order to convert light into electricity, we need to convert photons into electrons, the reverse of the process that generates light in the semiconductor laser. Not unreasonably, then, we again look first toward the p–n junction, and think about how we might reverse the lasing action.

In the semiconductor laser, a stream of electrons, in the form of an injection current, is pumped into a p–n junction. As a result, in the region where the p and n materials meet, there is a continuous recombination of electrons (from the n region) with holes (from the p region), thus producing photons whose energy (and therefore wavelength) corresponds to the band gap of the host semiconductor.

Suppose now, however, that we take the p–n junction and try instead to pull the electrons and the holes apart, away from each other, in the region where the materials meet. This can be done by applying across the junction a voltage that has the opposite sign to the one we used to inject current; that is, the p-region is now negative and the n-region is positive (Figure 5.1a). This is called a reverse-bias voltage, and it will have a value, in this case, in the region of 5–10v. The effect of this voltage is to form what is called a depletion region close to the junction (Figure 5.1b), where any holes and electrons have already combined, and there are no other charge carriers, of either sign, around (hence the term *depletion region*). The material in this region has a full lower energy band and an empty upper band (Figure 5.1c), and is depleted of electric charge.

Suppose further that a photon falls into this depletion region from the outside world. Provided that it has an energy greater than the band gap, this photon can be absorbed by the material. In so being, it can raise an electron across the gap from the lower to the upper level, putting an electron in the upper band and, of course, leaving a hole in the lower band: the photon has created an "electron-hole pair" (Figure 5.1d). There is now a negative charge and a positive charge in the depletion region, and both come under the action of the reverse-bias voltage that seeks to separate them. Hence, provided the electron can reach the safety of the n-region, and the hole the safety of the p-region, before they have a chance to recombine, they together represent a flow of electric charge through the voltage-driven circuit, and thus comprise a flow of current (Figure 5.1e). This current flow will be continuous if there is a continuous flow of photons into the depletion region; that is, if there is a continuous optical signal falling into the depletion region.

This simple picture tells us all the important things about semiconductor photodetection. First, because each photon produces (ideally, and at most) one electron-hole pair, the rate of charge creation (i.e., the electric current) must be proportional to the rate of arrival of the photons (the optical power). Hence, to maximize the sensitivity we must ensure that as much of the incoming optical power as possible can interact with the depletion region to create the electron-hole pairs.

Second, when the electron-hole pair has been created, the charges need to be swept very quickly to their respective safe havens, both to minimize their chances to recombine with each other and to ensure that the resulting electric current responds quickly to any changes in the optical power. It is this speed of response that determines the detection bandwidth. This

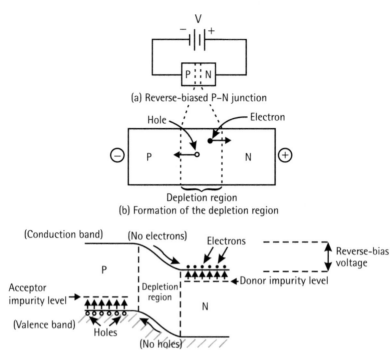

(a) Reverse-biased P–N junction

(b) Formation of the depletion region

(c) Energy-level diagram for the P–N junction

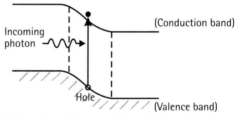

(d) Formation of electron-hole pair by incoming photon

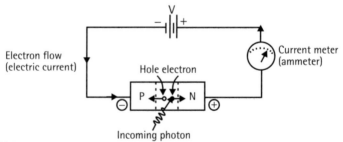

(e) Generation of measurable electric current from input light to depletion region

Figure 5.1 The P–N junction photodetector.

means that the depletion layer should not be too thick, and that the reverse-bias voltage should be large enough to pull the charges quickly across it.

The basic structure of a p–n junction photodetector, known as a p–n photodiode, is shown in Figure 5.2a. Clearly, the materials of which it is made must possess a band gap appropriate to the wavelength of the light that is to be detected.

The p and n semiconductor materials are joined as thin layers. The light to be detected enters from the top, passing right through the thin n layer with little interaction. The cross-sectional area of the depletion layer between the p and n layers will be of about the same size as the incoming optical beam, and thus can intercept a large fraction of the incoming optical power. The fraction of the optical power which is actually absorbed by the depletion layer (thus creating electron-hole pairs) now depends upon two things: the thickness of the layer and the wavelength of the light. In broad terms, for good absorption the thickness of the layer must be greater

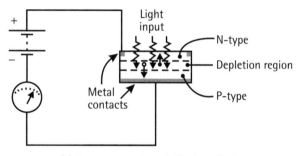

(a) Arrangement for a P–N photodiode

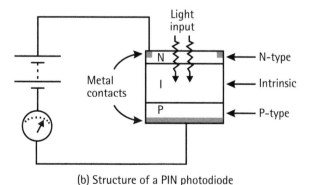

(b) Structure of a PIN photodiode

Figure 5.2 Photodiode structures.

than the wavelength of the light. The thickness of the depletion layer in a reverse-biased p–n junction is of order 1 μm, so wavelengths less than this will be well absorbed, but wavelengths greater than this, including our preferred ones of 1.3 μm and 1.55 μm, will not. Consequently, these longer wavelengths will not be sensitively detected by a p–n photodiode.

To some extent, the thickness of the depletion layer can be increased by increasing the reverse-bias voltage. This has the effect of pulling the static electrons and holes, in the n and p materials, farther apart. But there is a limit to the voltage that can be applied before the junction actually breaks down, and this limit is reached before the depletion layer thickness becomes large enough to detect these longer wavelengths sensitively. And, of course, the increased thickness also increases the response time, thus reducing the bandwidth of the device. This variation in bias voltage can, however, be used to trade off sensitivity and response time for the lower wavelengths.

The solution of the problem for the longer wavelengths is to interpose yet another layer of semiconductor material, 3–5 μm thick, between the p and n materials (see Figure 5.2b). This is a layer of undoped semiconductor material, or "intrinsic" semiconductor material, and it acts as an artificial depletion region. The structure now takes the form p-$i$ntrinsic-n, or PIN. The resulting device is known as a PIN photodiode.

The major advantage of this structure is that we now have the thickness of the depletion region entirely under control. We can, in particular, make it thick enough to ensure that most of the incoming light is absorbed by it. We must be careful not to make it too thick, for then the charges created by the photon would take too long to cross it, and the response time (and therefore the bandwidth) of the device would suffer. However, we now have an extra degree of control that allows us to trade off sensitivity and bandwidth for the longer wavelengths by careful adjustment of the depletion width and the bias voltage.

We can now optimize the device for any given set of system requirements. The PIN photodiode is very widely used in optical-fiber communications systems. A typical practical arrangement is shown in Figure 5.3. Again, this looks a little complicated, but each feature has a straightforward purpose, in view of what we have just learned about the action. The light emerges from the fiber, which is firmly attached to the detection photodiode by means of an optically transparent resin. The light then passes through the n$^+$-type indium phosphide (InP; the "+" simply means that the doping is much larger than usual, thus increasing the conductivity), a

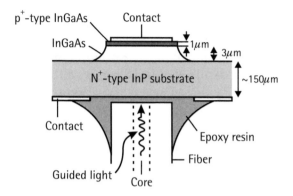

Figure 5.3 A practical PIN photodiode design.

material which is transparent to wavelengths in the range 1.3 μm to 1.6 μm, owing to the fact that its band gap is too large for photons in this range to create any electron-hole pairs. The light then enters the all-important intrinsic depletion region (in this case indium gallium arsenide [InGaAs]), which does have an appropriately small band gap for pair creation. This layer is wide enough (at about 3–5 μm) to intercept all the light from the fiber, but thin enough, in association with the reverse-bias voltage applied by the metallic contacts (shown in white) to allow rapid passage of the electrons and holes to their respective "safe" regions. The safe region for the holes is clearly the p$^+$-doped InGaAs near the top of the diagram, while that for the electrons is the n$^+$-doped material at the base. Right at the top is the metal contact that must form the negative voltage terminal; it is negative in order to attract the positive holes.

In this device, there is a 90% probability that any photon reaching the depletion layer of the InGaAs will give rise to an electron-hole pair. The device generates about half a microampère of electric current for a microwatt of input optical power (i.e., around half an ampère per watt, but we are usually dealing in millionths of these units, or less), at optical wavelengths in the range 1.3 μm to 1.6 μm. It responds to changes of power in about 100 picoseconds, giving it a bandwidth of about 5 GHz (PIN photodiodes with bandwidths of up to 50 GHz have been designed, although, to achieve these speeds, great care has to be taken with the associated electronics that follow the device). The PIN device is very compact, having a sensitive area with dimensions of order 5 μm $\times$ 3 μm, again very readily compatible with the dimensions of the monomode fiber core. The solid-state device will thus be compact, rugged, and durable; it requires only a low operating voltage (5–10v) and it has good temperature stability.

Clearly, this is a very useful device for our optical communications purposes, especially since, like the laser diode, it can again benefit from a well-established and highly-developed semiconductor technology.

5.3 The avalanche photodiode

There is another type of photodiode that also is in common use in optical communications systems and, with what we have already learned about PIN photodiodes, this is a device whose action is quite easy to understand: it is essentially a PIN photodiode operated in a different way.

Suppose that we take a PIN photodiode and apply to it a much larger reverse-bias voltage than usual, up to 10 times larger, at around 50–100v (rather than 10v). In this case there is now a much stronger pulling-apart effect on the electron-hole pairs when they are created in the depletion (intrinsic) region. So great, indeed, is this force, that the charges are accelerated to an extent where they themselves gain sufficient energy to create other electron-hole pairs, by colliding with other electrons and knocking them from the lower band to the upper band, much as the incoming photons do. These secondary pairs that are thus created can themselves create tertiary pairs, and so on, until there is an "avalanche" of charge creation, and thus a much larger current flow in the detection circuit. This avalanche process effectively gives us an amplification of the signal: unlike in the ordinary PIN photodiode, many more electrons and holes are produced than there are photons coming in (Figure 5.4).

This type of device is called an avalanche photodiode (APD) and it is very often used in optical communications systems. APDs are clearly more sensitive in terms of the current that comes out for the light power that goes in (by a factor of 50–100) compared with a PIN photodiode, and they are almost as fast. They do have a disadvantage, however, in that the actual number of avalanche charges produced by the original electron-hole pair is not a fixed number, but varies, since it is a statistical process. There is an element of hit-or-miss about it; a given charge may or may not gain sufficient energy to create another charge pair. The consequence of this is that the output current varies from one input photon to the next and thus from one moment to the next. An unwanted variation of signal of this sort is, as we already know, called noise, in analogy with unwanted sound, and noise

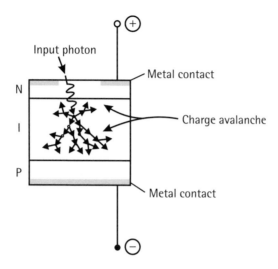

Figure 5.4 Action of an avalanche photodiode (APD).

is an important aspect of photodetection (indeed *any* detection) processes. This is a complex subject but its basic elements are straightforward, as we will see below.

5.4 Noise

To an optical communications systems designer, noise means unwanted variations of current in the output from an optical detector: there are many sources of such variations. It is noise that, more than any other factor, determines the limits to the performance of the system, a fact to which we have already referred, in a different context (Section 1.8).

We know that the loss present in the system, for example as a result of absorption and scattering of the light in the optical fiber, will cause the light power to diminish with distance. This results in a power level, at the output end of the fiber, that is much lower than was injected by the laser diode. There will be a length of fiber beyond which the power level is too low for the detector. Why? How low is too low? The answer is noise. When the signal level is down to about the same as the noise level, the signal level is probably too low, depending, to some extent, on the nature of the signal and the nature of the noise (Figure 5.5).

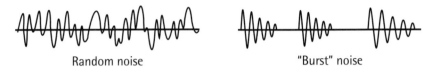

Random noise   "Burst" noise

Figure 5.5 Two types of noise.

There are two main kinds of noise in an optical communications system. The first is a result of the fact that the optical signal consists of a stream of photons. This stream consists of individual photons, so that we find that if we take a particular time interval, say one microsecond, a certain number of photons will arrive, but within the following microsecond, a slightly different number will arrive, and a different number again for the third microsecond. The average number over, say, any one full second, will be quite accurately the same (for a constant optical power) from second to second, but if we look at the signal on a small enough timescale, quite significant variations in level can occur (Figure 5.6). Of course, when we are dealing with high-speed optical communications signals, we do expect the detector to look at the signal over small time intervals, nanoseconds or even smaller, because these are the kinds of time intervals, between pulses, with which we are dealing. Hence this variation will be quite marked. The variation thus comprises a noise, and it is known as photon noise (sometimes quantum noise). The variation in the arrival rate of the photons is translated into a variation in the production rate of the electron-hole pairs and thence a variation in the flow of electrons; that is, a variation in the detector current. When looking at this type of noise on the current signal it is known as "shot" noise, and there was a brief discussion of this in Section 1.8. Shot noise was first studied in relation to the manufacture of lead shot for Victorian shotguns. This is done by allowing molten lead to drop from the top of a tall tower (a shot tower), up to 30m high. As the lead falls, it solidifies into balls, as a result of surface tension, all about the same

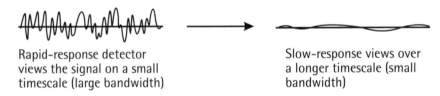

Rapid-response detector
views the signal on a small
timescale (large bandwidth)

Slow-response views over
a longer timescale (small
bandwidth)

Figure 5.6 A noise signal viewed by different detector bandwidths.

size, and suitable for use in shotguns. These balls are collected at the bottom of the tower, in a metal bucket, into which they can be heard to clatter with a certain type of audible noise, as a result of variations in their arrival rate (Figure 5.7). As it is with lead shot in a tower, so it is with electrons in a wire, and photons in a beam of light.

The second main type of noise in our systems is a direct result of the fact that electronic components, such as resistors, have a temperature—in other words, they have thermal energy. This, again as we know from the last chapter, takes the form of a constant jostling of the atoms or molecules of which the component is made: they are in constant motion, colliding with each other and with any electrons that are flowing through the material. This never ending, random jostling imposes itself on the motion of the electrons, so that there is a consequent random variation in the flow of current: a noise. This is known as thermal noise (or Johnson noise, after the man who first quantified it, in 1928). Of course, this thermal noise can be reduced by cooling the component, but that is expensive and usually quite inconvenient.

We have already seen that, in addition to shot and thermal noise, there is also avalanche noise produced by the charge-creation process in an APD. There are other (usually) less important sources of noise also. All of these types of noise have different characters that need to be taken carefully into account in order to get the last bit of performance from a system, but their broad restrictions are clear. Let us take a more detailed look at some of these restrictions.

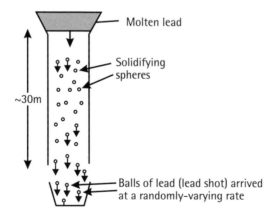

Figure 5.7 The shot tower.

In Chapter 3 we looked at the effects of fiber attenuation and dispersion on the propagating waves and pulses. Both effects will change the signals: the attenuation reduces their size; the dispersion changes their shape. Both effects alter their relationship to the noise.

Take, first, the example of the analog signal in the presence of noise as shown in Figure 5.8a. The analog signal is discernible but has been severely corrupted. If it were a speech signal, for example, it would sound mushy, and would quite often be unintelligible. The signal has been allowed to fall to too low a level in relation to the noise, and a system redesign would be required. The "signal-to-noise ratio (SNR)" has become too low.

Now take the case of a digital signal (Figure 5.8b). We know that this is a stream of pulses, the information being encoded as the presence or absence of pulses in particular time slots (yes or no). The optical receiver's job now is only that of determining whether a pulse is present or absent in the successive time slots, and this is rather easier than having to determine the actual level of the signal, as in the case of analog transmission (in

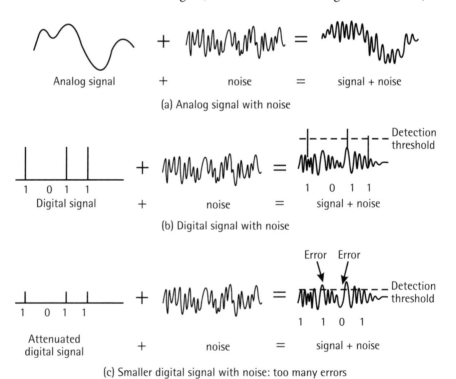

Figure 5.8   Signals and noise.

Figure 5.8a). As a result, the digital signal is more resistant to noise, which is one of the main reasons for using it.

In Figure 5.8b the noise level is about the same as before, but the digital pulses are still readily recognizable and, with a threshold as shown in the diagram, all the pulses and absence of pulses would be recognized, and the signal would be accurately reconstructed. In Figure 5.8c, however, we see that the pulses, although they would be still recognizable in the absence of noise, are just too small in its presence; no matter where the threshold level is set, the noise causes too many errors in recognizing the pulses. The SNR is again too low, so again a system redesign is required: for example, the length of the fiber might be reduced. The system is attenuation-limited.

Consider, now, the situation shown in Figure 5.9. Here, dispersion has broadened the pulses, but they still have a good size compared with the noise. In Figure 5.9a they have broadened but, even in the presence of the noise, they still have the effect of giving satisfactory yes or no answers. In Figure 5.9b, however, the reverse is the case. Even though the signal level is well in excess of the noise, the broadening has caused so much overlap between them that the noise prevents them from being separately distinguishable from each other. Here again the error rate would be too high. This system is now dispersion limited, and again a redesign would be needed: either the fiber length must be reduced, or the information rate must be reduced, so that there is a greater interval between the pulses (or some combination of the two).

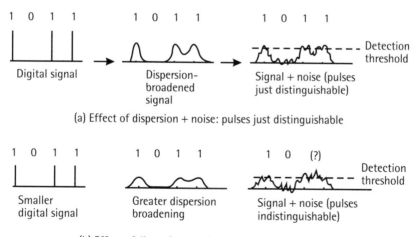

(a) Effect of dispersion + noise: pulses just distinguishable

(b) Effect of dispersion + noise: pulses indistinguishable

Figure 5.9 Effect on digital signals of dispersion with noise.

From these examples we may begin to see the important role played by noise in system design and how important, therefore, it is to try to minimize its effects. We can also see that noise, signal level, and bandwidth (information rate) all interact with each other. There is, in fact, a very definite relationship between these quantities, first derived by Claude Shannon in 1948 and presented, for interest, in Appendix C.

5.5   PINs or APDs?

Is it better to use PIN photodiodes or avalanche photodiodes (APDs) in our optical communications systems? There is no straight answer to this. We must again use that irritatingly vague phrase: *it depends on the system.* The detailed considerations are quite technical, so we must restrict ourselves to the bare essentials. The output signal from a PIN is at a relatively low level. In order to get an answer to the yes or no questions, the signal must be passed into a piece of electronics (a threshold detector) that will have its own sources of noise (all electronic systems have). So it is necessary to amplify the signal from the PIN before sending it on. The trouble is that this amplifier will itself add noise. It must, clearly, be a very fast amplifier (as fast as the signal it is amplifying, if it is not going to distort it), and the faster it is (i.e., the higher the bandwidth it has) the noisier it will be: the higher bandwidth will allow in the fast random variations as well as the slower ones. This is another example of the sensitivity-versus-bandwidth problem that applies to individual electronic circuits as much as it does to communications systems in general (we looked at this feature in Chapter 2); indeed, it applies to all physical systems.

Clearly, this amplifier must be very carefully designed and constructed so as not to degrade the signal too much: and this, of course, costs money.

The APD, on the other hand, provides a much larger output signal that probably will not need amplification before passing to the threshold detector. However, as was noted when we were describing its action, the avalanche process has some randomness associated with it and this itself adds noise to the signal, so that its output is inherently noisier than that of the PIN. So PIN + amplifier, or APD?

Both types of device have much the same optical characteristics, since they have the same basic structure and are made from the same materials (InGaAs for 1.3 μm or 1.55 μm optical wavelengths) and there really is little to choose between them for low- to medium-bandwidth systems. For the high- and very-high-bandwidth systems, the choice depends upon the more detailed requirements of the system, such as allowable error rate, installation environment, and cost. Designs and performances of both devices are improving constantly, however, and the best guess is that the APD ultimately will prevail, since it avoids the complication and expense of an extra amplifier.

5.6 Detection demodulation

Demodulation is the reverse of the modulation process that, as will be remembered from Chapter 1, is the process whereby signal information is impressed upon a carrier wave. The big advantage of this is that many carriers, with different frequencies, could be used, and because these frequencies could be recognized separately by the detector (by turning the tuning knob on a radio receiver, for example), many signals could be sent over the same channel (Figure 1.17).

In the case of radio and microwave frequencies, up to 10 GHz (i.e., 10^{10} Hz) say, the signal is impressed upon the carrier by modifying some aspect of the carrier wave (e.g., its amplitude, frequency, or phase) and the receiver selects that carrier frequency electronically when it wishes to receive it. This can be done because electronic circuits can operate up to about 10 GHz. After selecting the required carrier, its signal modulation (which will consist of a set of much lower frequencies) will be removed (in other words, the carrier will be demodulated) again electronically, and passed on for whatever use is to be made of it (see, again, Figure 1.17).

In the case of optical frequencies, things are rather different, for electronic circuits cannot operate at frequencies of 10^{14} Hz. The only physical systems that can respond sufficiently quickly to provide electric currents from optical waves are atoms and molecules. This is not surprising, since it is these that produced them in the first place. We have seen what can happen when an optical wave (i.e., a stream of photons) impinges upon atoms in a solid material. In the PIN photodetector, for example, the electrons are raised from the lower band to the upper band, giving rise to a current that is a measure of the optical power in the wave. Hence, if the information is in the form of a modulation of the wave power (equivalently, the wave amplitude), the demodulation is now automatically performed by the atoms themselves (Figure 5.10). The only problem is that the atoms in a solid are not as highly selective as is a tuned electronic circuit, because, as we

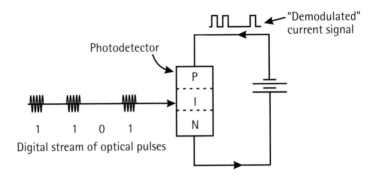

Figure 5.10 Demodulation by a photodiode.

know, the energy levels are in the form of broad bands, so a quite broad range of optical frequencies is capable of producing electron hole-pairs in any given photodetector material. However, the incoming photons must always have sufficient energy to bridge the energy band-gap, so there is frequency selectivity to this extent, at least.

There are several ways around this selectivity problem. For analog systems, one can make use of the very wide bandwidth of the optical system to modulate the optical wave with other carriers at microwave or radio frequencies. These are known as subcarriers, and they will each carry their own signal modulations as if they are separate carriers in radio channels. The current from the photodetector will now contain all of these modulated carriers, which can then be selected by conventional microwave or radio receivers, and demodulated as before (Figure 5.11).

In digital systems, the signal consists of a stream of pulses, and we can use a rather different strategy, as discussed in Section 1.6. Again making

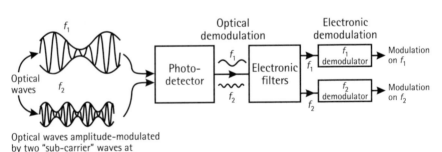

Figure 5.11 Sub-carrier demodulation.

use of the very wide bandwidth of the optical signal, we are able to interlace streams of pulses from various sources. Selectivity at the receiver is now performed on the digital output from the photodetector; for example, pulses can be directed according to the order in which they arrive, if this is how they have been interlaced at the transmitter (see Figure 1.10). Alternatively, if each signal source is allowed to put a burst of pulses on to the optical carrier, and there is an identifying pulse sequence at the front end of the burst, the demodulation now occurs by digital identification of each burst, and then appropriate redirection.

Finally, there is the case of the modern systems known as wavelength-division-multiplexed (WDM) systems. WDM systems increase the bandwidth-distance product for a given fiber by sending many different optical frequencies simultaneously down the fiber, each frequency (wavelength) comprising an entirely separate communication channel. For demodulation of the signals on these channels, it is first necessary to select the particular optical frequencies (wavelengths) before they reach any detector. This can be done conveniently by again using the wave interference effects that were described in Chapter 3, in relation to propagation in fibers. Figure 3.8 illustrated how an oil film could deflect different colors in different directions, and this also is the basis of wavelength selection in multiwavelength communications channels. Of course, we don't use oil films: the interference devices are very carefully constructed for the task at hand, but the basic principles are the same. We shall look more closely at WDM and its associated processes in Chapters 6 and 7.

5.7 Summary

In this chapter we have studied the photodetection process: the process whereby an optical wave, a stream of photons, is converted into an electric current. Only when the stream is in the latter form are we able to use the wonders of modern electronics to extract, manipulate, and use the information that has been impressed upon it. The photodetector plus its associated electronics together comprise the optical communications receiver.

We have seen how the photodetection process depends upon the properties of solid materials, in particular the way in which, in solids, the atomic energy levels organize themselves into bands. This structure allows incoming photons to kick electrons up from a lower to an upper band, and hence create electron-hole pairs that can then give rise to a measurable electric current in the circuit. By carefully constructing photodetectors from suitable materials, in suitable geometries, it has been possible to provide photodetection devices that

- Respond sensitively to the preferred optical communications wavelengths of 1.3 μm and 1.55 μm;

- Are sufficiently fast for wide-bandwidth optical communications systems;

- Consume only small amounts of electrical power;

- Are compact, rugged, reliable, and cheap.

We have also noted the special way in which optical signals are demodulated, and the crucial importance of noise in this process.

Having dealt, over the last three chapters, with the processes of light generation, modulation, propagation in fibers, photodetection, and demodulation, we are now in a position to put it all together, and consider some basic system designs and networks.

6

System Design

6.1 Introduction

In the preceding chapters we considered the main, separate features of optical communications systems: the nature of information transfer, the light sources, the impression of information on to the light in the modulation process, the guiding of the light in optical fibers, the detection of the light when it emerges from the far end of the fibers, and the extraction of the information in the demodulation process.

It is now time to put all these features together, in order to see how they interrelate to one another when trying to design a complete communications system. Of course, there are many different kinds of systems, depending on the particular requirements, ranging from a small amount of information to be transferred over a short distance, to a large amount over a long distance, and we shall need to put emphasis on different aspects of the design according to the demands. In doing so, it will become clear that some other components will be required, such as couplers, switches, compensators, and amplifiers.

Additionally, we will have to consider the problem of arranging that the many transmitters can communicate with many receivers, using an

127

integrated system (a national telephone system, for example), so we will consider some of the ideas in what is known as networking. This is a topic that is presently receiving a great deal of attention from telecommunications companies, as global interconnection requirements continue to grow very rapidly.

This chapter is concerned with complete systems, and there is a system tool, not yet described, that we will find very useful in system studies. The tool is the decibel (dB) and we shall first take a short digression to come to grips with this.

6.2 The decibel

Suppose that we have two lengths of optical fiber, A and B. Suppose now that we launch light separately into these fibers, finding that A reduces the optical power by a factor of 100, and B by a factor of 1,000. If, now, A and B are placed end to end and light travels first through A and then through B, the total loss of power clearly will be by a factor that is the product of the two loss factors; that is,

$$100 \times 1,000 = 100,000$$

We could also write this as

$$10^2 \times 10^3 = 10^5$$

Note carefully the values of the powers to which the number 10 has been raised in each case. These numbers are called exponents. We see that, for these exponents:

$$2 + 3 = 5$$

They simply add together.

Take another case. Suppose that the attenuations of the two fibers were much smaller, factors of 16 and 64, say. For the two together we should have a total loss factor of

$$16 \times 64 = 1,024$$

which can be written:

$$2^4 \times 2^6 = 2^{10}$$

The same rule is seen to apply also for these exponents, this time for the base number 2. We see that it might be quite convenient to use a system where the loss factor is always expressed as the power to which a given number (the base number) must be raised. For then, when considering successive losses in sequence, for different sections of fiber or for any other components, all we have to do is add together these exponents to obtain the total loss factor, provided, of course, we know the base number that is being used. The number chosen for the base number in matters of attenuation is 10. In our first example, the exponents 2 and 3 are said to be the logarithms to the base 10 of the numbers 100 and 1,000, respectively. This can be expressed in the shorthand form:

$$\log_{10}(100) = 2$$
$$\log_{10}(1000) = 3$$

Of course, any number can be expressed as a power of 10. Take, for example, the number 8.4:

$$8.4 = 10^{0.92}$$

So that

$$\log_{10}(8.4) = 0.92$$

We are using a logarithmic scale. (The old-fashioned slide rule worked on exactly the same principle.)

Now, for convenience, the exponents (i.e., the powers to which the base number is raised) are expressed in terms of a unit that is given the name bel, after Alexander Graham Bell, the inventor of the telephone. In fact, one bel is rather too large a unit for convenient practical use, so we use a submultiple, the decibel (dB), with 10 dB equal to 1 bel. The fibers A and B would thus possess attenuations of 20 dB and 30 dB, respectively. Fiber A followed by fiber B gives now a total attenuation of 20 dB + 30 dB = 50 dB. Clearly, if we wish to know the actual factor by which the power is reduced in total, we look at the 50 dB figure and reverse the process; that is,

$$50 \text{ dB} = 5 \text{ bels}$$

Hence the required factor is

$$10^5 = 100,000$$

10^5 is now the "antilogarithm" of 50 dB.

This system is universally used throughout the whole of telecommunications technology. For example, we know that the loss in an optical fiber increases with the length of the fiber. We can, in fact, specify the fiber loss in terms of decibels per unit length, usually expressed as dB.km^{-1}. Once we know this loss figure for a given fiber (it is a property of the fiber resulting from the absorption and scattering mechanisms discussed in Section 3.4), we know the loss factor for any given length of fiber. Let us take an example of a fiber with a loss of 2 dB.km^{-1}. What is its loss over 10 km? The answer is, of course, 20 dB, which is equal to 2 bels, equivalent to a factor of 10^2 (i.e., 100). Normally, it is not necessary to revert to the actual loss factors. All calculations can be performed in dBs. We shall see shortly how this can be done, and how useful it is.

The reader may be wondering about the connection with loudness of sound, which is also expressed in decibels. The acoustic dB performs a very similar function to the telecommunications dB. The reason for its usefulness in acoustics is that the ear operates logarithmically; that is, sound power levels in the ratios, of, say, 10:100:1,000 will have apparent loudness in the ratios 1:2:3. By using this logarithmic system the ear is capable of an enormous dynamic range, with the loudest perceptible sounds (before damage) being about 10^{12} (i.e., 1,000,000,000,000) times as powerful as the quietest sounds.

6.3 A simple analog system

One of the system advantages of optical-fiber communications links is that optical fibers do not suffer from the problem of environmental "pickup." In conventional, copper-based systems, any electric, magnetic, or electromagnetic fields, external to the system, which the copper conductors may encounter between the transmitter and the receiver, are capable of inducing electric currents in the conductors. These lead to pickup noise in the receiver; this reduces the signal-to-noise ratio (SNR) and hence degrades system performance. Steps must, therefore, be taken to exclude these fields by metal "shielding" of the conductors, adding to the bulk and the expense of the installation.

The optical fiber is made from silica, an insulator, and hence it does not suffer from induced electric currents. This freedom from interference induced by external fields is a considerable extra advantage for all types of optical-fiber communications system, short-distance, medium-distance, and trunk. But it is, of course, especially important when the communication environment is known to be noisy, such as in military applications or short-distance links in ships, aircraft, spacecraft, et cetera. In these latter applications the freedom from interference may well be much more important than the large bandwidth offered by the fiber.

We begin our system design studies with the simplest system of all—a short-distance, low-bandwidth analog link. Such a link might be needed to transmit, for example, a TV signal between two buildings, or sensor information across a few hundred meters in a temporary field trial. A fiber link could well be the preferred solution to such a problem, owing to an abundance of interfering noise, perhaps, or a number of metal obstructions, which would make it difficult to use either copper wires or short-distance radio communications, for the reasons outlined in the previous paragraph.

A primary feature of such a link is that there are no serious problems of attenuation or dispersion—the distance is too small for these to arise. Really the only problem is that of ensuring that the signal to be transmitted is recovered with sufficient accuracy within the allocated cost, the latter being quite small and, therefore, limiting.

The design of such a system, while technically relatively undemanding, does, nevertheless, provide an adequate introduction to some important features of system design, and this will be useful for the more demanding discussion that follows.

Let us consider some general points.

The signal, being analog in form, will consist of a continuously varying voltage, with some range of amplitudes determined by the device that has generated it (e.g., a TV camera or a measurement sensor). Since the distance for transmission is small, it will not be necessary to take advantage of the signal-to-noise benefits of digitization, and we can transmit it as an analog signal (and hence keep the costs down). However, the signal will need to be conditioned (i.e., adjusted in amplitude and bias level, and filtered to remove any unwanted components) before it has the right form for feeding into the particular light modulator that we are intending to use in our system.

Next, it is clear that the choices of light source, fiber, and detector are not critical, since the attenuation, dispersion, and speed of response are not demanding for these short distances, low bandwidths, and hence small bandwidth-distance products. Nevertheless, we do still need to minimize costs.

We would probably choose an arrangement such as is shown in Figure 6.1. The analog signal is conditioned, fed directly into an LED, which then feeds its light into a multimode fiber. The light emerging from the far end of the fiber feeds into a PIN photodiode for detection and demodulation. The demodulated signal (i.e., a replica of the original signal) is then amplified and suitably processed for whatever its intended purpose is. All of these components, the LED, the multimode fiber, and the PIN photodiode, are fairly cheap, and quite capable of performing these relatively straightforward tasks.

Even for this system there are some points that need careful consideration, however. These are probably best illustrated by looking at a specific set of requirements. Suppose that these are as follows:

1. Analog bandwidth: 10 MHz;

2. Transmission distance: 300 meters;

3. Maximum analog signal amplitude: 100 mV;

4. Signal recovery accuracy: better than 2% (this means that the recovered signal should not anywhere differ from the original signal by more than 2%);

5. Cost: less than $1,500.

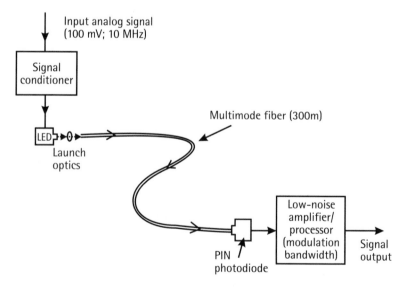

Figure 6.1 Schematic for a short-distance, low-bandwidth analog fiber link.

Since the choice of fiber is uncritical, the important design decisions to be made concern the light source and the photodetector. The arguments concerning these might proceed broadly as follows.

6.3.1 The Light Source

An LED can easily respond quickly enough to handle a 10 MHz analog signal. LEDs at a center wavelength of 0.85 μm are fairly cheap, so we tentatively choose this as a source at this stage, with its output power level yet to be decided.

An LED can only really feed into a multimode fiber, since its optical emission is quite widely divergent, and a monomode core is too small to accept much of it; moreover, a multimode fiber is usually cheaper. Let us choose a multimode fiber with an attenuation of 2.5 dB.km^{-1} at 0.85 μm and a bandwidth-distance product of around 10 MHz.km. Because this is a multimode fiber, this last figure is the result of modal dispersion, and it therefore does not depend upon the source's spread of wavelengths (as it would for chromatic dispersion). Hence, again, the LED (which has quite a large wavelength spread) is perfectly acceptable. Clearly, the 10 MHz.km figure is easily large enough for our present requirement, which is 10 MHz $\times$ 0.3 km = 3 MHz.km. With an attenuation of 2.5 dB.km^{-1} the attenuation over the transmission distance of 300 meters will be 2.5 $\times$ 0.3 = 0.83 dB, equivalent to a power loss of about 20% (i.e., $10^{0.083}$ = 1.21; the power is, therefore, reduced by a factor of 1.21, hence a loss of about 20%). However, only about 5% of the LED's light power will launch into even the multimode core. This is a result of the much poorer collimation of the light compared with the semiconductor laser diode. This, therefore, comprises a 95% power loss and is a major loss feature.

Since we are limited in cost, we will need to modulate the LED source directly, feeding a modulation current into the device. Clearly, we need as large an optical signal as possible from the LED, and the size of the signal we can get is limited by the linear range of the output-power-versus-input-current curve (known as its "characteristic"). For, if the signal goes beyond the linear limits (see Figure 6.2), the optical signal will be distorted by the curvature of the characteristic, and the 2% limit on the accuracy of regeneration of the signal will not be met. To be safe, we should ensure that the output optical power follows the input optical current to an accuracy of 1%. All of this means that the analog signal voltage (maximum 100 mV) must be adjusted so that it can cause a maximum swing of current of ±50 mA around a 100 mA set bias point (see Figure 6.2). This will require an

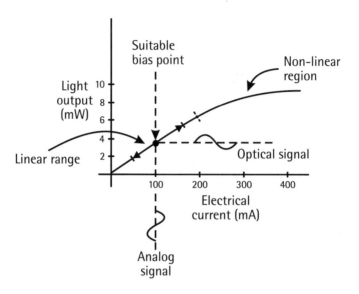

Figure 6.2 Bias point on the LED characteristic.

amplification of the signal (probably by a factor of about 10), a matching of electric impedances, and provision of a bias current. In other words, this is the conditioning of the signal that is required before feeding it to the LED.

6.3.2 The Photodetector

The light, when it emerges from the far end of the fiber, falls on to a PIN photodetector. A silicon PIN device is suitable for operation at $0.85\,\mu$m, is easily capable of handling 10 MHz of analog bandwidth, and is both cheap and readily available. It has a responsivity, at our chosen wavelength, of about 0.5 amp per watt of received optical power.

We need now to consider how much power the source LED should deliver, and this depends critically on the performance of the detector. The PIN photodiode converts the optical power fed to it by the fiber into an electrical current (0.5 amp per watt of optical power). This current will contain shot noise (see Section 5.4). The current must be converted into a voltage suitable for whatever function the output signal is to perform (e.g., feed into a TV receiver, or into a computer for sensor data analysis). So the current must be passed to a resistor (to convert it to a voltage) followed by a voltage amplifier, both of which will add more (mainly thermal) noise (again, see Section 5.4, if desired).

While a detailed noise analysis is beyond the scope of the present text, it is clear that the optical signal level feeding into the PIN photodiode must be large enough to allow the final output voltage from the amplifier to exceed the total noise level by a factor of at least 100. Then, the optical signal can be recovered to an accuracy of 1% which, with the 1% nonlinearity of the LED, keeps us within the overall 2% limit. The signal-to-noise requirements are eased by the fact that the noise levels increase with bandwidth, and our bandwidth here is quite small. However, there are the two sources of noise to be considered: shot noise and thermal noise. The thermal noise is introduced by the detector itself and its following amplifier: it is independent of the optical signal. This means that the optical signal must be about 1,000 times larger than its intrinsic shot noise before it enters the detector, in order to ensure that the output from the detector has an SNR of 100 after thermal noise has been added.

The detailed analysis in this case shows that the photodiode should deliver an average current of at least 50 nA to the resistor/amplifier in order to achieve the 2% accuracy. This means that the average input optical power to the photodiode (with its 0.5 ampere per watt responsivity) should be at least 100 nW. The loss in the fiber is 20%, so we shall need 125 nW to be launched into the fiber. Since only 5% of the LED output is launched into the fiber, this means an average of 2.5 μW should be emitted by the LED when set to its midrange bias point. Since the bias point corresponds to approximately half the maximum possible (thermally safe) output, this latter will be about 5 μW. We must now increase this by a factor of 6 (about 8 dB) as a safety margin to allow for a degradation of components over time; the fiber's alignment with the source or detector might change, for example, or the characteristics of the devices themselves might change. Hence, we end up with a final figure of 30 μW for the LED output.

6.3.3 System Summary

Our design is now complete and apparently satisfactory. To summarize its features:

1. Source: an LED with power output greater than 30 μW at a center wavelength of 0.85 μm;

2. Fiber: multimode; 2.5 dB.km^{-1} attenuation (at 0.85 μm); 10 MHz. km bandwidth-distance product;

3. Detector: silicon PIN photodiode with responsivity 0.5 ampere per watt;

4. Signal conditioning: analog signal amplified to provide a voltage signal to the LED, biased to the midpoint of its linear range and limited in amplitude to the extent of the linear range (±50mV);

5. An output low-noise, modulation-bandwidth amplifier, following the PIN photodiode, to provide a voltage signal appropriate for the intended task (TV or computer);

6. A total cost of less than $1,500, including the cost of assembly (although this might entail some shopping around).

We can begin to see now what the system designer's basic thought processes are. These may be summarized (and prioritized) as below (with the system's specifics in parentheses):

1. What are likely to be the limiting factors in the design? (Accuracy of regeneration and cost.)

2. Can an analog system be used, to reduce costs? (Yes.)

3. What is the cheapest type of source that can be used, and at what wavelength? (LED at $0.85\,\mu$m.)

4. What is the cheapest type of fiber that can be used with this source and what attenuation can be expected at that wavelength? (Multi-mode: 2.5 dB.km^{-1}.) Is this an acceptable attenuation figure? (Yes.)

5. What is the bandwidth-distance product for this fiber? (10 MHz.km.) Is this acceptable? (Yes.)

6. How should the signal be conditioned for accurate modulation? (Bias the LED at midpoint of a linear part of the characteristic, and limit the signal voltage swing to lie within the linear range.)

7. What is the best photodetector to use for this wavelength? (Silicon PIN photodiode.)

8. How much optical power should be fed to the photodiode in order to provide the required accuracy of output signal in the face of detector/amplifier noise? (100 nW.)

9. How much power must, therefore, the source be able to provide? (30 μW.)

10. Are all the required components available? (Yes.)

11. Is the total cost of the system within the prescribed limit? (Yes, it is hoped.)

Obviously, detailed prioritization and iterations will vary from system to system, but the general thought processes are quite well represented in this example.

From this relatively straightforward illustration we turn now to a more demanding one: a long-distance, high-bandwidth digital system. The design for this will be very different, as will the priorities.

6.4 A long-distance, high-bandwidth, digital system

Suppose we are now confronted with the following problem. Five analog television channels (BBC1, BBC2, ITV3, 4, and 5) are to be transmitted, simultaneously, over a distance of 50 km between two towns, because one of the towns has very poor reception from the terrestrial transmitter, owing to its geography. Each channel consists of an analog signal with a bandwidth of 5 MHz, and the quality of the received signal must be high enough to provide a TV picture essentially indistinguishable from that which it would ideally have received directly from the broadcasting transmitter. Costs must be kept to a minimum, since we are a company bidding for a contract to do the job, and there is a lot of competition.

The bandwidth and the distance are both quite large, but we are hoping to avoid having to use any repeaters in the link, since these will add considerably to the cost. The preliminary design argument might proceed as follows.

Over a distance of 50 km, both attenuation and dispersion will be important factors to be considered. The attenuation is likely to be quite large, and the received power therefore quite small. In order to recover a good quality TV picture, we will need a large SNR at the output. This will be difficult to achieve, not only as a result of the large attenuation, but also because the large bandwidth necessary in the receiver will allow in a large amount of noise (noise power increases with bandwidth).

It is fairly clear, then, that we will need a digital system, in order to utilize the SNR advantages of this type of system. (Recall that these advantages stem from the fact that the digital receiver has only to decide whether a pulse is present or absent in a given time slot, and this can be done with a lower SNR than would be needed for an analog system of the same quality.)

Having decided on a digital system, the first thing that we need to do is to determine the required digital bandwidth; that is, the required bit rate.

6.4.1 Digitization

To determine the bit rate, we must return to the ideas developed in Section 1.6, where the digitization process was described.

1. In accordance with the Nyquist requirement, each analog waveform is sampled at a rate equal to at least twice the bandwidth; that is, at 2×5 MHz $= 10^7$ samples per second. This is equivalent to sampling the waveform at 0.1 μs intervals (Figure 6.3).

2. Each sample amplitude is then converted into a digital code: the amplitude is represented, in binary code, by a number of bits, each of which can take on the value 0 or 1.

3. We require a high-quality signal, so let us choose to represent the sample amplitude by one of 512 levels, equivalent to a representation accuracy of about 0.2%. This requires 9 bits, since $2^9 = 512$.

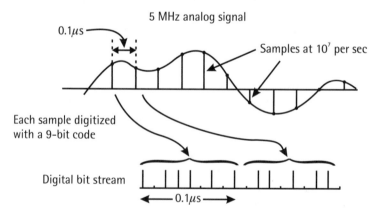

9 bits in each successive 0.1μs $\equiv$ 90 Mb.s^{-1} bit stream

Figure 6.3 Digitization of the analog signal.

4. There are now 10^7 samples every second, each sample being represented by 9 bits: a set of bits, each one of which is either a pulse or an absence of a pulse (i.e., 1 or 0). Hence there now will be 9×10^7 pulses per second, giving a bit stream of 90 Mb.s^{-1}.

5. There are five analog signals for transmission, and they must be transmitted simultaneously. This means that the bit streams from each of the channels must be interwoven, taking, perhaps, one pulse from each channel in turn (Figure 6.4). This process is known as multiplexing. In order to separate the bits appropriately at the receiver, there needs to be some labeling so that the decoder knows which bit belongs to which channel in any given sequence, and which channel is which. This will add, let us say (depending on the coding scheme), another three bits to each sample, making a total of $9 + 3 = 12$ bits per sample.

6. We now have $10^7 \times 12$ bits per channel, and five channels to be transmitted simultaneously, making a total bit rate of $5 \times 10^7 \times 12 = 600$ Mb.s^{-1}.

Our design problem, then, is to transmit this bit rate over 50 km of fiber.

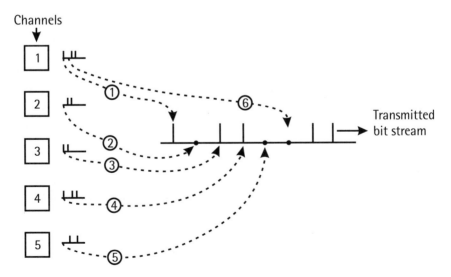

Figure 6.4 Interweaving of channel bit streams.

6.4.2 Bit-error rate and signal-to-noise ratio

As previously mentioned, with such a large digital bandwidth required at the receiver, one problem is the large noise level that will be present. Digitization reduces considerably the required SNR; this is why we use it. However, over these distances and with these bandwidths, a reasonably high value will still be required. A detailed noise analysis is complex and, again, beyond the present level of description, but the required SNR clearly must be such as to allow most 1s and 0s to be recognized correctly by the receiver. The performance of the digital receiver is thus normally specified by quoting a bit-error rate (BER); that is, the rate at which mistakes (interpreting a 1 as a 0 or vice versa) are made. An acceptable BER for a telecommunications system is usually taken as 10^{-9}; that is, there is only one mistake allowed in 10^9 bits received. This will ensure that the original bit stream is quite accurately reproduced. This means, in turn, that when the original signals are converted back to analog form, the analog signals will be an accurate production of their originals to the extent of one part in 512, which was the accuracy of the digitization process. This is good enough for high-quality TV pictures and it can be achieved with a much lower SNR than would be required if an analog system were being used. The actual value will be decided later, when we have considered other aspects of the design, in particular the choice of photodetector.

6.4.3 The fiber

We need to transmit a large bandwidth digital signal over a large distance. Clearly, the fiber we use must have low attenuation and low dispersion. It must, therefore, be a monomode fiber operating at one of its low-loss windows. Let us tentatively choose a wavelength of 1,300 nm (1.3 μm), which lies within the second low-loss window for silica fiber (see Section 3.4). The reason for this choice is that the fiber, sources, and detectors are all relatively cheap and easily available at this wavelength (when compared with 1,550 nm, for example). We hope to be able to use a single length of fiber (albeit with joined sections) without the use of repeaters, for repeaters are expensive to manufacture, to power, and to maintain. If a single unrepeatered length is not possible at this wavelength, we might be forced to use repeaters, or perhaps try another wavelength (probably 1,550 nm). Ultimately, the final decision will depend upon cost.

At 1,300 nm wavelength, monomode fiber is readily available with attenuation of 0.4 dB.km^{-1} and a dispersion figure of 3 ps (nm.km)$^{-1}$. Let us see if this will allow a suitable repeaterless system design.

This fiber will have a total attenuation, over the 50 km, of 50 × 0.4 = 20 dB. The dispersion will depend on the spectral width of the laser source, which we have not yet considered.

6.4.4 The receiver

The received signal level is, of course, crucial. The whole system design is geared to providing what the receiver needs in order to give us the required output.

Probably the most important component in an optical receiver is the first port of call, the photodetector. This, in a digital system, will be either a PIN photodiode or an avalanche photodiode (APD). Silicon-based devices will not perform well at 1,300 nm, so we turn to indium gallium arsenide (InGaAs), which will. The APD will need less input optical power than the PIN, but it is more expensive and requires much higher operating voltages, with associated safety implications. We decide, therefore, to do the calculations for both, to see what comes out for each.

First, we need to establish that each device is able to cope with the 600 Mb.s^{-1} bandwidth. This is, indeed, the case. InGaAs devices can operate easily up to about 2 Gb.s^{-1}. The responsivity of these devices is also quite good, at about 0.8 amp per watt, better than the silicon devices at their best. On performing the noise analysis, we find that, at 600 Mb.s^{-1} digital bandwidth, the PIN requires a peak pulse power of 0.5 μW for a 10^{-9} BER, while the APD requires only 0.05 μW.

How easy is it going to be to provide pulses with these powers?

6.4.5 The power budget

We now know how much power the receiver needs for a satisfactory signal for each of two possible photodetectors. How powerful a source is needed to provide these powers? To answer this question we must consider the various losses in the system: we must consider the "power budget."

The major loss is due to the 50 km of fiber, and we have already calculated that to be 50 × 0.4 = 20 dB. Another source of loss results from the fact that 50 km of fiber cannot be laid as a single section; it will be laid (probably) in 25 successive 2-km sections, which then have to be joined together. They will be joined in fusion splices, each of which will have an associated additional loss. A good fusion splice will introduce a loss of 0.05 − 0.08 dB, but we cannot be sure that they will all be good ones, so let us allow a safe value of 0.1 dB per splice. There will be 25 × 2 km sections in our 50 km, requiring 24 splices, so the total splice loss will be 24 × 0.1 = 2.4 dB.

The next loss to consider is that of launching the light from the semiconductor laser source into the fiber. The efficiency of launch is much better than for an LED, but it is still only about 60%, corresponding to a loss of 2.2 dB (i.e.,$10 \log_{10} 0.6$).

Correspondingly, there is a loss when the light strikes the photodiode; some will be reflected and some will not reach the active area; this adds, typically, another 1.5 dB of loss.

Finally, we must provide a system margin. This allows for general degradation of the system over time (the splices and source/detector junctions might misalign or corrode, for example) and for fiber bend losses. Usually about 8 dB is allowed for this.

So we may summarize our power budget as in Table 6.1.

(Note how easy it is to add the losses when they are in dBs. This is, of course, precisely why we use this logarithmic form.)

Armed with this figure, we are now in a position to find out if there is a suitable source that can overcome all these losses.

6.4.6 The light source

The first light source that comes to mind is a semiconductor laser operating at 1,300 nm: an InGaAs laser. We need, however, to do some calculations to be sure that this is suitable for our purposes.

The source is required to provide 0.5 μW (PIN) or 0.05 μW (APD) to the photodetector, after 34.1 dB of loss. This loss corresponds to a loss factor of 2.57×10^3 (i.e.,$10 \log_{10} [2.57 \times 10^3] = 34.1$).

Hence, for a PIN photodiode, the laser must have an output power of 1.28 mW (i.e., $2.57 \times 10^3 \times 0.5.10^{-6}$ W), while for the APD, the figure is 128 μW.

TABLE 6.1   The Power Budget

Loss Feature	Loss (dB)
Fiber	20.0
Splices	2.4
Launch	2.2
Detector	1.5
System margin	8.0
Total	34.1 dB

Both of these figures are well within the capability of commercially available semiconductor lasers at this wavelength, so we choose the higher value; let us say a laser with a safe (i.e., long life) minimum output power of 3 mW, in order to operate the receiver with the (cheaper) PIN photodiode.

One disadvantage of the higher power is that the laser might well have a larger spectral width. This will be about 2 nm, so we need now to look at how this interacts with the bandwidth requirements.

6.4.7 Dispersion

The chosen fiber's dispersion figure is 3 ps $(nm.km)^{-1}$. The laser spectral width of 2 nm and the transmission distance of 50 km now give a total system dispersion of $3 \times 2 \times 50 = 300$ ps $= 3 \times 10^{-10}$ s. For the 600 Mb.s^{-1} bit stream, the spacing between the bits is $1/(6 \times 10^8) = 1.67 \times 10^{-9}$ s. We can allow the bits to spread, as a result of dispersion, by about half this period, or $\sim 0.8 \times 10^{-9}$ s. The actual dispersion we have calculated is comfortably less than this (300 ps $= 0.3 \times 10^{-9}$ s), so all is well.

Another way of expressing all of this is by way of the bandwidth-distance product. For 2 nm spectral width, the dispersion of the fiber is 6ps.km^{-1}. The bandwidth-distance product is the reciprocal of this; that is, 167 Gb.s^{-1}.km. Allowing only half of this for our half-period spread condition reduces this to 83.5 Gbs^{-1}.km. This is well able to contain our requirement of 0.6 Gb.s^{-1} $\times$ 50 km = 30 Gb.s^{-1}.km.

6.4.8 Signal Conditioning and Coding

All that remains now is to consider some of the associated electronics. The analog signals are first digitized using an analog-to-digital converter (ADC), and then interwoven/coded to feed the resulting pulse stream into a voltage amplifier. The amplifier provides an output pulse stream of the correct amplitude for feeding directly into the laser, allowing the electrical pulses to be converted into pulses of optical output with peak values of around 1.3 mW. At the same time, the correct current bias is provided for the laser. This will be just above the value that causes it to lase (the threshold value), so that the delay in the onset of lasing action (i.e., the time for the buildup of the stimulated emission) does not slow down its response to the input pulse stream (see Figure 6.5).

At the receiver, the PIN photodiode is followed by an amplifier and then a circuit that decides whether a pulse is present or absent in given time slots, according to whether the signal is above or below a preset threshold in these slots. This circuit then provides the information that

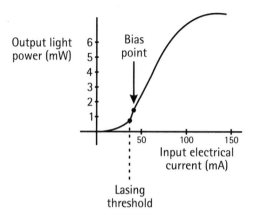

Figure 6.5 Semiconductor laser bias point.

sends a cleaned-up version of the pulse stream to the demultiplexer, which separates the pulses from the five different channels. Each channel pulse stream is then fed into a digital-to-analog converter (DAC), which regenerates the original form of the analog signals for feeding into the television receiver.

6.4.9 System summary

The full system is illustrated in Figure 6.6. Its main specifications are summarized below.

1. Source: InGaAs semiconductor laser, center wavelength 1,300 nm, linewidth 2 nm, minimum (for long life) output power 3 mW;

2. Fiber: Monomode fiber, loss (at 1,300 nm) 0.4 dB/km^{-1}, dispersion figure 3 ps (nm.km)$^{-1}$;

3. Receiver: InGaAs PIN photodiode, followed by a voltage amplifier, decision circuit, and demultiplexer.

4. Performance: Digital system operating at 600 Mb.s^{-1} over 50 km with a BER of 10^{-9}, a total loss of 34.1 dB, and a total dispersion of 3×10^{-10} s. This allows the five TV channels to be transmitted simultaneously (i.e., multiplexed) over the transmission distance with the required reception quality.

So again we have a system satisfactorily meeting the requirements.

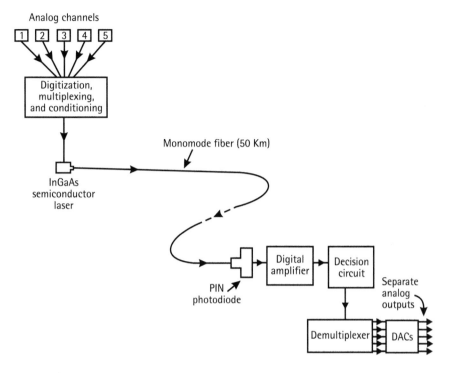

Figure 6.6 The complete long-distance, high-bandwidth digital communications system.

Note the following differences in design approach when compared with the previous (analog) system.

1. The distance and bandwidth are much higher, so it was necessary to choose a type of system that is more tolerant of noise and dispersion (a digital system using monomode fiber).

2. It was necessary also to choose a wavelength at which fiber loss is very low (1,300 nm).

3. The choice of wavelength dictates the material to be used for the source and the detector (InGaAs).

4. The acceptable BER determines the source power required in the face of total system losses (1.28 mW).

5. The bandwidth required determines an allowable source spectral width in the face of fiber dispersion (2 nm).

We might have found that some of the numbers did not fit, in which case a rethink would have been required. This is very often the case in system design.

Note that this system is an example of one that is attenuation-limited. The limitation to performance is due to attenuation rather than to dispersion: the system remains well within the dispersion limit and could approximately double in length before this limit is reached.

We next consider an example of a system where the reverse is the case.

6.5 Trunk systems

Of course, it is sometimes necessary to design systems where the transmission distances are much greater than 50 km. For the transatlantic crossing, for example, the distance is about 5,000 km. Such systems are often referred to as trunk systems. How should we tackle this problem? We will not here get into another detailed design, but it will be valuable to become acquainted with the main ideas for these systems. What needs now to change for distances of several thousand kilometers?

The first obvious change to make to the system we designed for 50 km (in Section 6.4) is to use fiber with an even lower loss. This means moving to an operating wavelength of 1,550 nm, where the loss comes down to about 0.2 dB/km^{-1} in monomode fiber. This roughly doubles the distance possible, for a given bandwidth, with the 1,300 nm fiber (0.4 dB/km^{-1}), but sources and detectors are more expensive at the higher, 1,550-nm wavelength. Furthermore, the optical dispersion has to be considered more carefully now. Fibers can be made with very low dispersion at 1,550 nm, but they are special fibers and quite expensive. They are called dispersion-shifted fibers (DSF) and they were described in Section 3.4.2.

In any case, we are still here only considering distances of about 100 km and, at several thousand kilometers, we are now discussing much greater distances than that. For these greater distances the answer is to include repeaters at 75 km to 100 km intervals. For analog links these would be straightforward amplifying repeaters (Figure 6.7), where the optical signal is detected, the resulting electronic signal is amplified, and the amplified signal then remodulates another laser for continued transmission. An

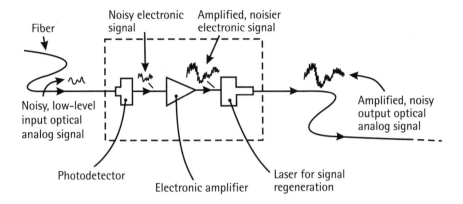

Figure 6.7 An analog repeater.

important disadvantage of this procedure is that each successive repeater will introduce noise, so that the noise accumulates along the transmission path, producing an upper limit, on the fiber length, of about 500 km.

In fact, very few long-distance links are analog in form, for reasons that have already been discussed; they are almost all digital systems. Hence, we should consider a digital repeater.

The digital repeater is fundamentally different from the analog repeater in that the pulse stream can be detected and then fully regenerated, rather than amplified (Figure 6.8). It is often, indeed, called a regenerative repeater. The important advantage is that the noise does not accumulate, in this case, since once an input pulse has been recognized, its associated noise is irrelevant and is effectively rejected. Some noise will be associated with the regeneration process, but this will be the same at each repeater, the same at the last in the line as at the first; there is, therefore, no noise accumulation. This helps enormously, and now distances of 5,000 km or more can be spanned. Of course, each repeater has to be electrically powered and maintained, and this is particularly troublesome for submarine cables (under the Atlantic Ocean, for example), where the repeaters are powered separately via copper wires in the cable, and the repeaters need to be ultrareliable to minimize the frequency of maintenance. All of this, of course, costs a lot of money for, on a transatlantic link, about 50 such repeaters might be required.

Furthermore, the repeaters also become exceptionally costly and relatively unreliable for bit rates in excess of about 1Gb.s^{-1}, for the electronics

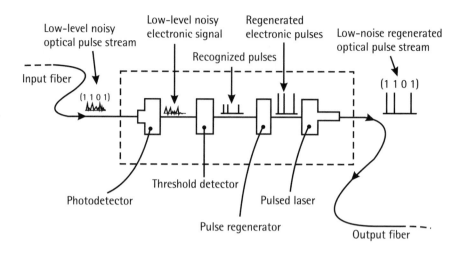

Figure 6.8 A digital regenerative repeater.

become quite tricky at such speeds. These speeds are not, of course, large in the context of optical communications systems. Hence we must look toward the methods that allow an increased bit rate for repeatered trunk systems. The first stratagem is to increase the number of fibers in the cable, and transatlantic cables (for example) might contain as many as 10 separate fibers, 5 for each direction of transmission. However, each fiber will require its own repeaters, so costs and failure rates soon become unacceptable.

A better solution is to use an optical-fiber amplifier. This is a section of optical fiber that is doped with a rare-earth element such as erbium or praseodymium. The atoms of the dopant can be pumped into a state of population inversion, as was described with respect to laser action in Section 4.3. The population inversion in the dopant material exists between two levels whose energy difference corresponds to the energy of a photon at the optical wavelength that is to be amplified. We noted, in Section 4.3, that this meant that any photon entering such a system would lead to many more such photons being produced, as a result of the process of stimulated emission. This was the essence of laser action, but we also noted there that it can also be used in optical amplification. An optical-fiber amplifier based on these ideas is illustrated in Figure 6.9. Note that it needs an optical coupler for the pump light. This amplifier is described more fully in the next chapter.

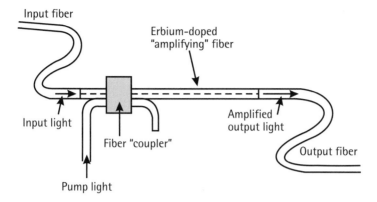

Input fiber

Erbium-doped
"amplifying" fiber

Input light

Amplified
output light

Fiber "coupler"

Output fiber

Pump light

Figure 6.9 An optical-fiber amplifier.

It is clear that this is a very convenient form of amplifier for an optical-fiber communications system, since it becomes an in-line amplifier, with the input and output fibers being simply fusion-spliced on to the ends of the amplifying fiber. This arrangement is geometrically very convenient, and results in very little insertion loss. In such an amplifier, the optical signal is amplified optically, not electronically, so no detector, electronic amplifier, or relaunch laser is involved; the signal remains always at the optical level. This avoids the use of the costly and vulnerable electronics needed for the regenerative repeaters, but it also has two much more important advantages.

First, the modulation bandwidth is no longer restricted by the electronics: the complete optical input is now amplified, together with its modulation signal, almost no matter what is its impressed modulation bit-rate.

Second, the optical-fiber amplifier has a fairly large optical bandwidth of about 30 nm (Figure 6.10), equivalent to about 4 THz ($4 \cdot 10^{12}$ Hz). This means that more than one wavelength can be amplified at the same time, and thus leads to the possibility of the fiber carrying many independent channels of information, each at a different wavelength, at the same time. This is a technique known as wavelength-division multiplexing (WDM), and it is a hot topic in this technology at the present time (2000). With narrow-spectral-width lasers, up to about a hundred channels at wavelength spacings of about 0.3 nm can be accommodated by such an optical amplifier. This means that, if each channel has a digital modulation bandwidth of 10 Gb.s^{-1}, say, the total fiber modulation bandwidth capacity is 1,000 Gb.s^{-1}. But since each channel can be detected separately, the

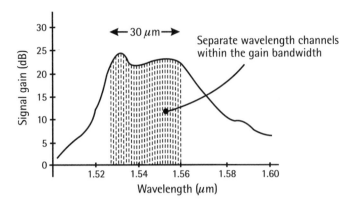

Figure 6.10 Optical bandwidth for an erbium-doped fiber amplifier.

sources, detectors, and other components need only operate at 10 Gb.s^{-1}. This is proving to be an extremely cost-effective way of increasing the bandwidth of single fibers, and is especially useful for upgrading fiber systems that have already been installed. We will take a more detailed look at WDM in the next chapter.

Of course, there are problems with the use of optical amplifiers. No valuable advance is made without difficulty. One of these problems is that there is now some noise accumulation, since the digital signals are not regenerated at each amplifier as they were in the regenerative repeater. However, with careful design, this accumulated noise can be made acceptably small. Another, more important, problem is that of dispersion. The amplifier will compensate for the effects of the fiber attenuation but not for its dispersion. If the operating wavelength is 1,550 nm, as it almost certainly will be, in order to take advantage of the very low attenuation at that wavelength, we can use DSF. This fiber is designed to give zero dispersion at 1,550 nm by allowing the chromatic and waveguide dispersion effects to cancel at 1,550 nm (see Section 3.4.2). But there will still be some dispersion owing to the nonzero linewidth of the source, and so we need to try to compensate for this over very long distances. The systems are now dispersion-limited. Dispersion compensation is another hot topic in technology today, and we will look at this also in the next chapter.

We have met, in this section, some ideas that are very important for modern trunk systems. The solutions to the problems that are presented by them will define the coming generation of new fiber communications systems.

We will return to all of these matters in the next chapter, which will interest readers wanting to acquaint themselves with the more advanced aspects of the subject.

6.6 Networks

Any telecommunications system needs more than simply the means for connecting a transmitter and a receiver. It requires the means by which many "customers" can communicate with one another, each customer being able both to transmit and receive information. In order to arrange that this can be done, systems must be organized into networks, which were mentioned briefly in Chapter 1. Let us now consider some more detailed aspects of this subject.

Two possible configurations (or "topologies") are shown in Figure 6.11. In the first (Figure 6.11a), the star network, any customer ("node") can communicate with any other via the central hub. The hub contains the control, which can either take the form of passive switching, or it can convert the optical signals to electronic ones and amplify or regenerate them, much as do the repeaters in the trunk systems discussed in the previous section. This is active switching.

In the second case (Figure 6.11b), each transmitting customer links to a ring, which then circulates the information until it reaches the receiving customer, who taps it off. Clearly, the information must contain a customer "address" that can be recognized by the intended receiver. Control is effected from outside the ring. If the information is not intended for any given receiver, the signal is regenerated by that receiver and passed on. The system may be designed to tolerate the failure of, say, 10% of the receiving nodes. One of the advantages of the ring topology is that information can flow in either direction around the ring, so that if there is a break somewhere in the ring, the system does not fail.

Many combinations of these basic topologies are possible, including a "star of stars" (see again Figure 1.12b), or a star/ring, depending on the requirements. Such arrangements are common for what are known as local area networks (LANs), where the information transfer is restricted to a large building, or a university campus, for example. Trunk networks are arranged rather differently. There is, of course, the main trunk connection (between cities or countries, for example), and then dispersal of the information at the trunk terminal (Figure 6.12). The local dispersal will cascade

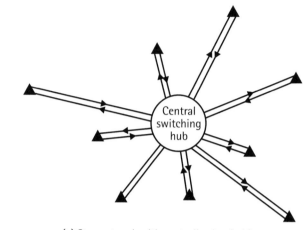

(a) Star network with centralized switching

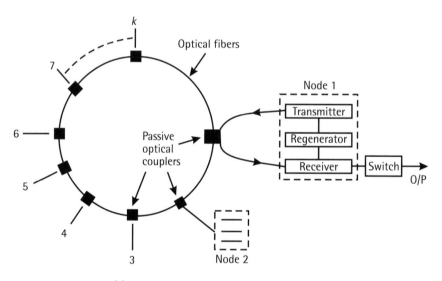

(b) Ring network with passive optical couplers

Figure 6.11 Optical-fiber communications networks.

down a hierarchy of networks depending on the bit rates and the number of customers at the different levels. It may, for example, be dispersed first to a set of stars using optical fibers, and then via copper cables or wires to the various nodes, or there may be first a digital optical dispersal followed by an analog optical dispersal, followed by copper wires (Figure 6.12). All will depend upon bit rates and numbers, and all ultimately determined by cost.

It is clear from this brief discussion of networking that yet more optical components are required. We need means by which optical fibers can add their signals to other fibers and the means by which this process can be controlled. We need optical couplers and switches. These, also, will be considered in the next chapter.

6.7 Summary

In this chapter we have brought together many of the ideas developed in the preceding chapters, to understand how optical-fiber communications systems are designed. These systems are, of course, the ultimate goal for all of optical-fiber communications technology.

We have noted how the various features need to be traded off against each other in order to meet a particular specification within cost constraints. We have also noted that systems make further demands on the technology by requiring additional components, such as:

- Couplers;

- Switches;

- Multiplexers;

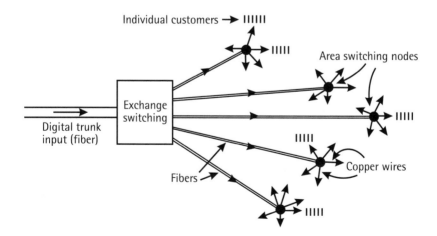

Figure 6.12 Network hierarchy.

- Optical amplifiers;

- Dispersion compensators;

- Special fibers.

Chapter 7 is intended for those readers who would like to know something of how these components work. Some other, more advanced, ideas will also be included.

Although this next chapter is more advanced than the rest of the book, for the most part it contains only a more sophisticated discussion of ideas that have already been explained and should, therefore, be accessible to those whose patience is not yet exhausted.

7

Advanced Topics

7.1 Introduction

In preceding chapters we looked at the principles upon which optical-fiber communications systems work. We demonstrated the need for certain components and defined the features that presently limit system performance.

In this chapter we will study the design of some of the components needed in modern optical systems, and take a brief glimpse at some of the work that is going on, at the front edge of the technology, to improve system performance still further. These developments are taking us toward the next generation of systems.

Few new principles will be introduced in this chapter: we will for the most part be dealing with topics that can be understood on the broad basis of the principles that have already been explained. However, some of the ideas involved here are quite sophisticated applications of these principles, giving further evidence of their power and importance.

This chapter will deal, essentially, with ways in which light can be manipulated—in direction, in frequency, in amplitude, and in dispersion.

The greater the control we have over the light, the better the system design will be.

We begin with the control of direction.

7.2 Splitters, couplers, and switches

It is frequently a requirement to pass light from one fiber into another fiber. This process is referred to as the "coupling" of light from one fiber to another. Examples where this is required are the coupling of many optical signals at different wavelengths into one fiber, in wavelength division multiplexing (WDM); the coupling from a "transmit" fiber into a "receive" fiber in the hub of a star network, or into the circulating ring of a ring network; the coupling of pump light into the amplifying fiber of a fiber amplifier or fiber laser. How can any of these be achieved?

Consider, first, the arrangement shown in Figure 7.1b. Light in the core of a fiber (1) is fed into a planar waveguide that splits into two, as shown. A planar waveguide operates on the same principles as the cylindrical optical fiber. It differs from the fiber in having its core laid into a rectangular slab of material, the core now being a strip of material doped to have a slightly higher refractive index than the material of the slab, so that guiding by total internal reflection can take place, just as in a fiber. The reason for using a slab is that the guiding path geometry can easily be tailored to be anything we wish, since it is only a question of injecting the doping material from the top of the slab in the required pattern (Figure 7.1a), which needs only to be a few centimeters long. Thus we can readily form the Y pattern shown in Figure 7.1b. When the light from the fiber 1 encounters the fork in the waveguide, it will split into two, and the planar waveguides then feed into two fibers (2 and 3) which are joined to their ends. So we now have half of the original light power in the input single fiber feeding into the two output fibers, 2 and 3; we have a "beam splitter." This device is also sometimes known as a 3dB coupler (since $10^{-0.3} = 1/2$).

In fact, its simplicity is deceptive and such a device has to be quite carefully designed, in order to ensure that each of the output fibers has the same amount of light, and to ensure that not too much light is lost, totally, in the splitting process. The design thus has to make sure that the splitting junction is not too abrupt, and that the speed of the light in the planar guide is about the same as that in the fiber core; otherwise, again, light will be lost from the guide. This device can just as easily work in the reverse direction,

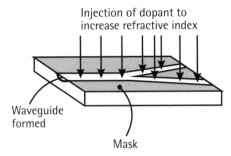

(a) Formation of a planar waveguide structure

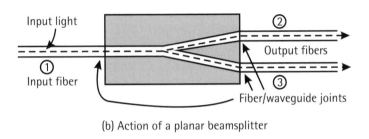

(b) Action of a planar beamsplitter

Figure 7.1 A planar waveguide beam splitter.

so that light fed into the two fibers 2 and 3 can be combined into fiber 1. We now have a beam combiner that will allow two independently generated signals to be passed down a single fiber for transmission, as is required in a WDM system, for example, where a cascaded array of such devices might be used (see again Figure 6.11).

Consider now a different pattern of planar guide, as shown in Figure 7.2. Here again there is one input port and two output ports, but this time there are two parallel planar guides. These planar guides are close enough for light in one to "leak" into the other. This leaking process might seem to violate the principle of total internal reflection at each waveguide boundary, but it is a consequence of the fact that the boundary looks quite fuzzy at the level of the atoms of the material on either side of the boundary. The light can only sense the position of the boundary to an accuracy of the order of its wavelength, so that if the second waveguide is only about a wavelength away, the light does not see a sharp boundary between the two waveguides, but only a fuzzy barrier through which it can leak from one guide to the other (see Figure 7.2).

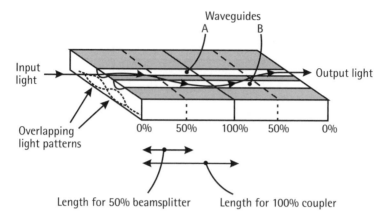

Figure 7.2 A planar waveguide coupler.

Suppose, then, that a fiber is glued to one end of one of the planar guides and light is launched into the guide from the fiber (guide A in Figure 7.2). As the light propagates in guide A it slowly leaks into guide B. After a certain distance, which will depend upon the separation of the guides, the optical wavelength, and the refractive index of the material between the guides, all of the light will leak from A to B. When B contains all the light, it will start to leak back to A, since the barrier is quite symmetrical. And so it will continue, with the light leaking back and forth between the two guides. If now we design the guide structure—in relation to the wavelength of the light, of course—such that all the light has leaked from A to B just as guide B comes to an end (and feeds into its end fiber), then we have coupled all the light into fiber 2 (see Figure 7.2). If, on the other hand, we shorten the length of the structure so that only half the light has coupled from A to B when it comes to an end, then fibers 1 and 2 receive equal amounts of light, and we have another beam splitter. This beam splitter might be about 2 cm long, while the full coupler would be about twice as long, depending, of course, on how close the guides are. (Such couplers are often called directional couplers, since light in the second guide always propagates in the same direction as the first. It never couples to the reverse direction.)

Now, suppose that an electrical voltage is applied across the material between the two planar waveguides. This can be done by fixing strips of metal on either side of it to act as electrodes for application of the voltage from a power supply (Figure 7.3). The effect of this is to change the

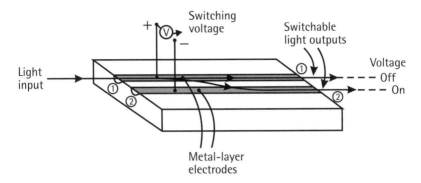

Figure 7.3 A planar waveguide switch.

refractive index of that material, because it restricts the motions of the atomic electrons and thus makes it more difficult for light to pass through it. The effect, in turn, of this, is to alter the ease with which light can couple between the two guides, A and B, and hence alter the coupling length over which all the light is leaked from one guide to the other. Hence, by controlling this, the voltage can determine how much light is passed to port 1 and how much to port 2. By suddenly changing the voltage from one level to another, for example, we could switch all the light from port 1 to port 2. We have an "opto-electronic switch," and this clearly would be very useful in switching optical telecommunications signals in the hub of a star network, for example. Such switches can operate very rapidly, in about 1ns (10^{-9}s). They are best when fabricated from crystalline materials whose refractive index is especially sensitive to applied voltage, owing to the ordered arrangement of their molecules. An ordered arrangement is more easily disturbed by outside influences.

It is also possible to design fiber couplers without any intervening planar guides. Figure 7.4 shows one arrangement that has been used: the "polished coupler." A fiber is glued into a curved groove within a silica block (see Figure 7.4a), which is then carefully polished until the surface is close to the fiber core, the cladding having been largely polished away. Two identical such blocks are then faced together (Figure 7.4b) so that the cores of the two fibers are close enough to allow leakage to take place, just as in the planar guide. In this way, light from one of the guides can couple directly into the other, the degree of coupling now being controlled by sliding the blocks over one another, in a carefully controlled way, along the direction of light propagation. Such a coupler is very useful in a

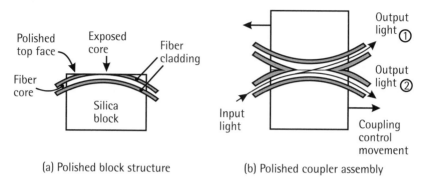

(a) Polished block structure   (b) Polished coupler assembly

Figure 7.4 The polished fiber coupler.

whole range of fiber-to-fiber coupling functions, pumping light into a fiber amplifier or laser, for example (Figures 7.5a and 7.5b).

A more recent development for the controlled switching of light is that of microelectromechanical systems (MEMS) or micro-opto-electronic-mechanical systems (MOEMS). These technologies are concerned with the fabrication of arrays of microminiature mirrors or movable waveguides, with individual dimensions of order 100 μm. Control is effected by mechanical movement over micron (10^{-6} m) distances, brought about by externally applied electric fields. Arrays of switches allowing the switching of, for example, 128 input fibers into any one of 128 output fibers, can be fabricated on a chip only 2.5 cm $\times$ 2.5 cm in size. The switching speed is relatively slow, of order 1 ms (10^{-3} s), but this is adequate for many telecommunications switching purposes. These switching operations are of low loss and consume little power. A great deal of research effort presently is being put into this technology.

Such are the stratagems used to split, combine, couple, and switch light between fibers; they are straightforward in principle but, owing to the fact that the distances involved must be controlled to within about one wavelength (i.e., of order 1 μm; 10^{-6} m), they are quite difficult to construct, and tend to be fairly expensive.

7.3 Optical–fiber amplification and lasing

In Section 6.5 reference was made to the optical-fiber amplifier. It was noted there that this device is very useful in providing in-line amplification

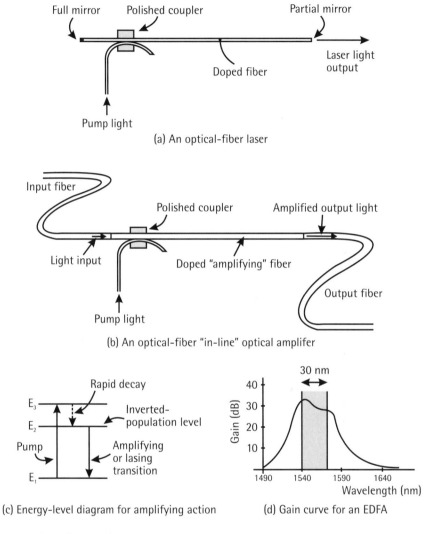

(a) An optical-fiber laser

(b) An optical-fiber "in-line" optical amplifer

(c) Energy-level diagram for amplifying action

(d) Gain curve for an EDFA

Figure 7.5 Optical-fiber laser and coupler.

of the optical signal at the optical level, for employment in repeaters for trunk communications systems. The particular advantage of this device was that it could amplify the optical signal over a broad wavelength band ($\sim$ 30 nm), so that it was capable of amplifying, simultaneously and wholly at the optical level, a complete set of multiplexed channels in a wavelength-division-multiplexed (WDM) system. With this device there was no need for any demultiplexing or electronic amplification, thus reducing both costs and technical difficulties considerably. Before dealing with WDM in more detail, in the next section, we shall first take a closer look at this optical amplifier.

The most important component in an optical-fiber amplifier is a length of several meters of silica fiber that is doped (to the extent of a few parts per million) with one of several possible rare-earth elements, depending on the operating wavelength. The rare-earth series of elements comprises a range of elements that have very similar atomic structures for their outermost electron energy levels. The elements differ among themselves only with respect to their atomic number (i.e., number of protons in the nucleus), and in the numbers of electrons in their inner energy levels, some of which are unfilled. This peculiar feature (of unfilled inner levels) makes them all very similar chemically, since it is the outer levels that normally take part in chemical reactions.

The advantage of all of this for optical amplification in fibers is that, when these atoms are doped into the random structure of the silica fiber, the inner levels are well shielded from that structure and thus remain relatively sharply defined in energy. This, in turn, means that they can be quite efficiently pumped into a population inversion, using the processes described in detail in Section 4.3.

Provided, now, that we choose an element that has the possibility of a pair of inverted-population energy levels that correspond to our operating optical wavelength, the population inversion can be used to amplify an incoming signal by the process of stimulated emission (also as explained in Section 4.3). Hence we can construct an optical-fiber amplifier using these principles. Furthermore, if the ends of the fiber are provided with mirrors, so that photons reaching the ends of the fiber are returned to enhance the stimulated emission, we can construct an optical-fiber laser.

The element erbium (Er) provides a pair of levels suitable for amplification at a wavelength of 1,550 nm. As a consequence, the erbium-doped fiber amplifier (EDFA) is becoming a very important component in trunk fiber communications systems, because the fiber attenuation is, of course, also at its lowest around this wavelength. The basic design of an EDFA is shown in Figure 7.5b.

By adding the end mirrors to the amplifying fiber, we can construct an erbium-doped fiber laser (EDFL), as shown in Figure 7.5a. This laser, while it has great promise as a source for optical-fiber communications systems, owing to its ready compatibility with the transmission fiber, it cannot yet compete successfully with the semiconductor laser owing to its extra bulk, cost, and vulnerability to shock. It has many uses, however, in other application areas.

Of course, both the amplifier and the laser must be pumped at a wavelength shorter than the operating wavelength, in order first to populate an upper energy level in the required way (see Figure 7.5c). This pump light is usually fed into the amplifying fiber via a polished fiber coupler of the sort described in the last section (see Figures 7.5a and b) at one end. The pump light then propagates coaxially with the incoming light, providing linear, progressive amplification along the fiber axis.

A typical gain curve (i.e., plot of amplification against wavelength) is shown in Figure 7.5d. From this variation it can be seen that there is a gain-bandwidth, where the gain exceeds 30 dB (i.e., greater than $\times$ 1,000), of about 30 nm. This is the range that is to prove so valuable in repeaters for WDM systems.

Let us now look more closely at WDM itself.

7.4 Wavelength–division multiplexing

The possibility of using wavelength-division multiplexing (WDM) for increasing the bandwidth capability of an optical fiber was mentioned in Chapter 6. It is worth looking at this technique in more detail, since it has a number of important advantages and it is the subject of a lot of attention at the front end of present-day technology.

WDM relies fundamentally on the fact that most natural physical effects remain fairly constant as the wavelength varies over a range that is large compared with the spectral width of lasers. For example, the attenuation spectrum of a silica fiber in the region of the 1,550-nm window is shown, in Figure 7.6a, in relation to a typical semiconductor linewidth of 1 nm. This corresponds to a frequency range of about 125 GHz. It is clear that up to about 30 such linewidths could be fitted into this window with a spacing of about 1 nm between them. Hence we can envisage having up to 30 telecommunications channels, with their center wavelengths at this spacing of 1 nm, all propagating independently in the fiber with perfectly acceptable attenuations. Clearly, this would increase the bandwidth of the link by a factor of 30. WDM comprises the technique by which this advantage is exploited.

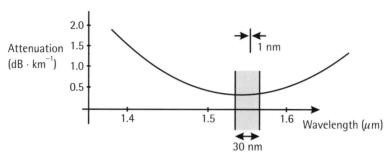

(a) Variation of attenuation in the region of 1,550 nm wavelength

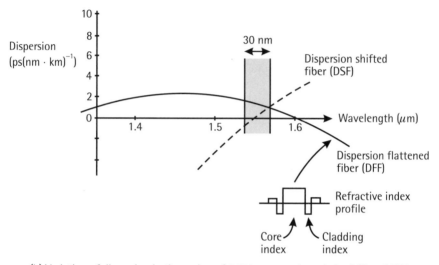

(b) Variation of dispersion in the region of 1,550 nm wavelength for DSF and DFF

Figure 7.6 Variations over the fiber-amplifier wavelength range.

But what happens to the dispersion? The variation of dispersion over this same wavelength range is quite small in dispersion-shifted fiber (DSF; see Figure 7.6b), but it can be reduced still further by the use of dispersion-flattened fiber (DFF). This fiber uses a more complicated refractive index profile to provide a waveguide effect that flattens the dispersion over the required range (see Figure 7.6b). Hence the dispersion remains within acceptable limits right across the 30-nm range.

However, the really important advantage of WDM lies in the fact that optical-fiber amplifiers provide good (and fairly constant) amplification over such a range also, something that was emphasized in Chapter 6. Hence, repeaters using these amplifiers can amplify this range of WDM wavelengths, together with the digital modulation signals they carry. This can be done without any electronics, and thus without any of the technical difficulties and expense involved in providing many repeaters, each with their many channels, and with each channel containing electronics having to be capable of operating at speeds in excess of 1 Gb.s^{-1}.

At the transmitting end, each separate channel must be "multiplexed" on to the single fiber, perhaps via an array of the fiber couplers shown in Figure 7.1, arranged as shown in Figure 7.7. At the receiving end, the many channels are first separated optically and then individually detected and decoded, thus avoiding a requirement for electronics at the full link bandwidth, providing instead a set of detectors and decoders each operating at the lower channel speeds (see Figure 7.7).

The optical separation can be effected in a variety of ways. One simple way is illustrated in Figure 7.8a, using a prism such as Isaac Newton used to split light into its component wavelengths. Each different wavelength is deflected through a different angle by the prism, so each channel in the WDM signal emerging from the trunk fiber will also be deflected through a different angle. This allows each channel to be directed toward its own photodetector for detection and subsequent processing. This is, in fact, a rather crude method for separating (demultiplexing) WDM channels, and better methods exist. These include the use of waveguide couplers (remember that the strength of the coupling depends upon wavelength; see Figure 7.8b) and interference filters (remember how the oil film, in Figure 3.8, was able to deflect differing wavelengths through different angles, to give rise to an interference pattern).

Clearly, these multiplexing and demultiplexing components must be carefully designed in order for them not to introduce too much loss, because with this, much of the WDM system advantage would be forfeited.

The thrust of the technology is now toward systems that can have as many as 100 channels, each with a 10 Gb.s^{-1} capability, giving a total bandwidth of 1,000 Gb.s^{-1}, or 1 terabit per second (1 Tb.s^{-1}). These systems use special lasers with a narrow wavelength spread (less than 0.1 nm), and are commonly referred to as dense wavelength-division-multiplexed (DWDM) systems. We are now entering the realm of terabit technology.

In WDM systems, the attenuation of the fiber is compensated by the optical amplifying repeaters in the link, at, say, 50-km spacings. However, the dispersion is not compensated; it accumulates. These systems are,

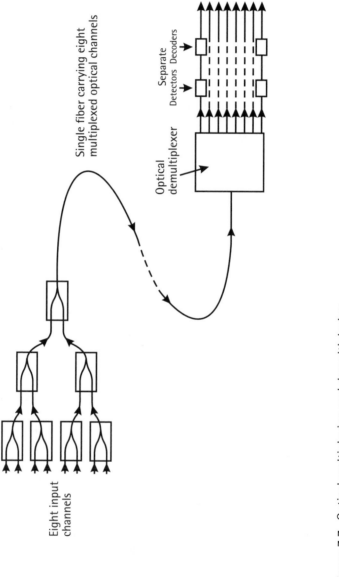

Figure 7.7 Optical multiplexing and demultiplexing.

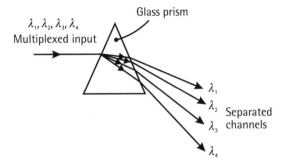

(a) Wavelength separation with a refracting prism

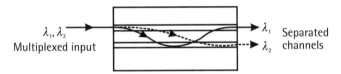

(b) Principle of waveguide separation of wavelengths

Figure 7.8 Optical demultiplexing principles.

therefore, usually dispersion-limited. If we are to improve performance still further, this problem has to be tackled. This will be our next topic for discussion.

7.5 Dispersion compensation

The use of optical amplifiers essentially overcomes the problem of attenuation in trunk telecommunications systems, because the loss of signal level along the fiber path is compensated by the repeater amplification. The dispersion is not compensated by the amplifier, however; this is a separate problem to be addressed.

One solution to this problem is to use the fact that in some fibers the dispersion is positive at a given wavelength (1,550 nm, say) while in others it is negative (see Figure 7.9). In other words, in some fibers the longer wavelengths of light travel faster than the shorter ones (positive dispersion), while in others the reverse is the case (negative dispersion). This is a consequence of the particular way in which the refractive index varies with wavelength for a given fiber material and waveguide structure. A

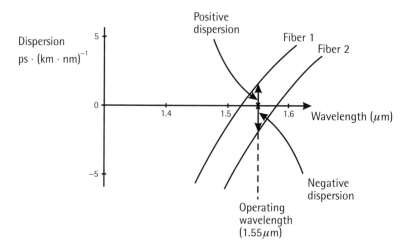

Figure 7.9 Fibers for dispersion management.

dispersion-compensation scheme thus springs immediately to mind: use one section of fiber with positive dispersion followed by an equal length of fiber with negative dispersion, and the two effects will cancel. In other words, the spreading of the pulses resulting from the longer wavelengths traveling faster than the shorter ones in the first section of fiber is turned into a compression of the pulses in the second fiber, where the longer wavelengths now travel slower than the shorter ones. The two effects cancel, and the pulses emerge much as they were when they entered the fiber (Figure 7.10).

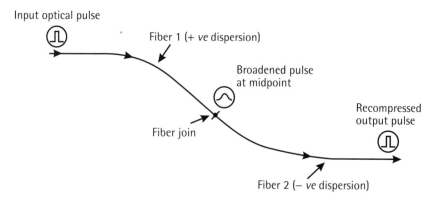

Figure 7.10 Dispersion management system.

Such arrangements are called dispersion-managed systems, and they can be used to some advantage. However, they are tricky to set up, because they require just the right type of fiber of just the right length, and impose limitations on link topology and on laser spectral widths. They have a part to play in dispersion compensation but are not the complete answer.

An alternative solution is to use a passive compensator. One form of passive compensator is the fiber Bragg grating (FBG). The FBG is an increasingly important component in modern optical-fiber communications systems, and we need, first, to understand how it works.

When a light wave strikes the boundary between two different materials, there is a reflected wave and a transmitted wave, as we saw in Section 3.2. This is a direct consequence of the fact that light travels at different speeds in the two materials, and the shock of a sudden change in velocity causes some light to be reflected. Suppose now that we have the material structure shown in Figure 7.11. This consists of successive layers of two materials, 1 and 2, with a structure 1-2-1. A wave striking the 1/2 boundary will be partially reflected, and partially transmitted on to the 2/1 boundary. At this second boundary it will again be partially reflected and partially transmitted (see Figure 7.11).

We now have two reflected waves. These waves will interfere, as waves always do when traveling along the same line. If the peaks and troughs each coincide (i.e., the waves are in phase) we have a wave of twice the amplitude of either one (Figure 7.11a). If the peaks coincide with troughs and vice versa, the two waves cancel, and there is no resultant wave (Figure 7.11b). And there are all the other possibilities between these two, giving resultant waves with amplitudes between zero and twice that of each wave. (All of these ideas were presented in Section 3.3.) It is clear from this argument that, for the maximum possible resultant, the distance of the double passage, from the first boundary to the second and back again, must contain a whole number of wavelengths (1 or 2 or 3, etc.). Only for those wavelengths for which this is true will there be a strong reflection. (It should be noted, however, that the wavelength of light is smaller in an optical material than it is in free space, by a factor equal to the refractive index of the medium. This is a result of the fact that the frequency is the same, because time does not slow down, but the speed is smaller, by a factor equal to the refractive index, and the product of frequency and wavelength must always be equal to the speed.)

Consider now the extended grating structure shown in Figure 7.12a. The distances between the boundaries are all the same, so that a wave

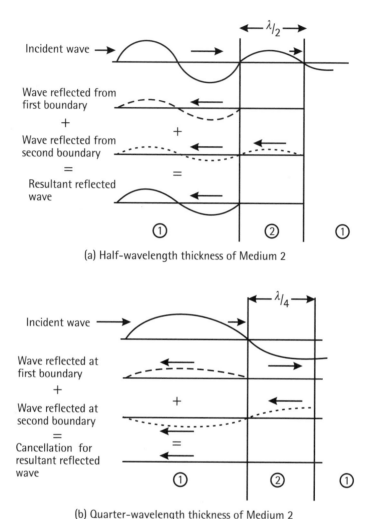

Figure 7.11 Principles of Bragg reflection.

whose wavelength is equal to twice this distance will be very strongly reflected by the structure. The longer the structure (i.e., the more reflecting elements it contains), the more highly selective in wavelength it will be. The reflection spectrum of a grating structure designed to reflect, very selectively, at a wavelength of 1,550 nm is shown in Figure 7.12b. The effect of reducing the number of elements in this grating is also shown.

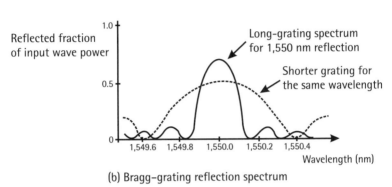

Figure 7.12 The Bragg grating.

These structures can be formed in optical fiber. They are formed by illuminating the fiber with ultraviolet (UV) light, using a mask with the required pattern (see Figure 7.13). Where the UV light strikes the fiber core, it increases the refractive index, via a photochemical reaction. The phenomenon is called photosensitivity. Thus we shall have successive sections of low and high refractive index in the fiber, with a prescribed separation, according to the wavelength that we wish to reflect in the fiber. Such is the structure known as the fiber Bragg grating (FBG), named after the physicist Lawrence Bragg, who used the layers of atoms in crystals to provide such reflective structures for X rays.

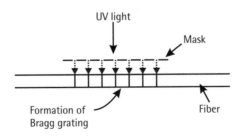

Figure 7.13 Fabrication of a fiber Bragg grating (FBG).

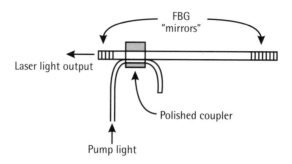

Figure 7.14 FBGs as mirrors in a fiber laser.

FBGs have, as a result of their high selectivity, many roles to play in fiber technology. For example, they can act as highly reflective mirrors in a fiber laser (Figure 7.14), selecting for reflection just the lasing wavelength. Clearly, the FBG at the output end of the laser should be shorter than at the other end, so that its reflectivity is not as great, and some light can emerge to be used. FBGs can also act as filters in a WDM system, providing the capability, via their wavelength selectivity, for separating the various wavelength channels at the receiver. Such filters can even be "tuned," to some extent, since control of the spacing can be effected either by stretching the fiber or by changing its temperature.

For our present purposes, in dispersion compensation, we use a special kind of FBG. This is illustrated in Figure 7.15a. The spacing between the reflective boundaries gets progressively smaller and smaller. This is known as a "chirped" grating, the name derived from the analogy of the continuously rising frequency of birdsong. Suppose now that a broad spectrum of waves falls on to this chirped FBG (Figure 7.15a). What happens now is that the longer wavelengths are reflected at the front of the grating and the shorter ones from the back, the wavelengths corresponding to the grating spacings in each case. Consequently, the long wavelengths arrive back at the front end of the grating before the shorter ones. This is, of course, just what we need for dispersion compensation. A pulse that has been broadened by negative dispersion has suffered from the longer wavelengths present traveling more slowly than the shorter ones. If this pulse is now passed into the chirped grating (properly designed for the dispersion that has occurred), then the dispersion can be reversed by the differential reflection (Figure 7.15a) of the grating. Clearly, if the broadening is the result of positive dispersion, the grating will need to be reversed, with the smaller grating spacings at the front end.

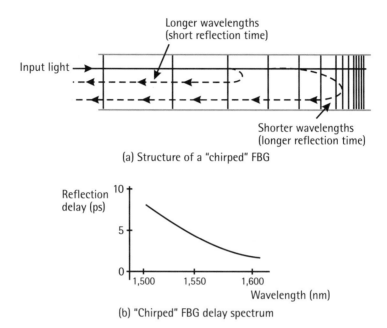

Longer wavelengths
(short reflection time)

Input light →

Shorter wavelengths
(longer reflection time)

(a) Structure of a "chirped" FBG

Reflection
delay (ps)

(b) "Chirped" FBG delay spectrum

Figure 7.15 The "chirped" FBG.

A scheme for effecting this dispersion compensation is shown in Figure 7.16. The scheme works well for the chromatic (i.e., material and waveguide) dispersion, although, as usual, care must be taken to ensure that the compensator does not introduce too much loss. Unfortunately, even when all of this chromatic dispersion has been compensated, there is yet another form of dispersion that cannot be compensated in this way. We shall now take a very brief look at this.

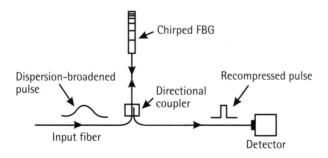

Chirped FBG

Dispersion-broadened
pulse

Recompressed pulse

Directional
coupler

Input fiber

Detector

Figure 7.16 Dispersion compensation with an FBG.

7.6 Polarization-mode dispersion

In Section 2.2 we learned that electromagnetic waves consist of electric and magnetic fields oscillating at right angles to the direction of travel. The direction of oscillation of the electric field, say, might be vertical or it might be horizontal (Figure 7.17). Suppose that the cross-section of the fiber core is not circular but elliptical (Figure 7.18a). If the direction of the electric field of the optical wave propagating in the fiber is aligned with the major

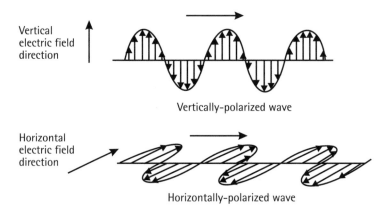

Vertical electric field direction

Vertically-polarized wave

Horizontal electric field direction

Horizontally-polarized wave

Figure 7.17 Linearly-polarized light waves.

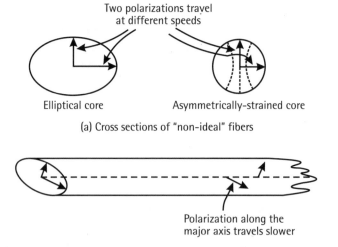

Two polarizations travel at different speeds

Elliptical core   Asymmetrically-strained core

(a) Cross sections of "non-ideal" fibers

Polarization along the major axis travels slower

(b) Delay between polarization components resulting from an elliptical core

Figure 7.18 Elements of polarization mode dispersion (PMD).

axis of the ellipse, it is not unreasonable to suppose that it will travel down the fiber with a speed different from that when it is aligned with the minor axis (Figure 7.18b). The waveguiding conditions clearly are different for the two cases. This is indeed what happens. The major-axis component travels slower than that along the minor axis. Slight deviations from perfect symmetry in either the fiber geometry, its material composition, or the strain in the material (see Figure 7.18a) mean that different directions of wave oscillation travel at different speeds. The direction of the wave oscillation is related to the polarization of the wave, and the differential speed effect is known as polarization-mode dispersion (PMD). It is difficult to control the polarization state of the light while it is traveling in the fiber, because bends, twists, external pressures on the fiber, and random variations in the core shape, composition, and strains all conspire to change it as the light progresses (Figure 7.19). The consequence is that the light's polarization state and its PMD also vary randomly, leading to a progressive, if rather small and quite slow, broadening of the pulse with distance, of about 1 $ps.km^{-\frac{1}{2}}$ (it only increases with the square root of distance, owing to its random nature). Although the effect is quite small, it is significant after chromatic dispersion has been reduced to an absolute minimum, and becomes the bandwidth-limiting factor when the bit rate exceeds about 100 $Gb.s^{-1}$, because at that rate the PMD broadening is becoming significant in relation to the width of the pulses. One obvious stratagem for reducing PMD is to try to make the fiber as symmetrical, in every respect, as possible. This means that greater care has to be taken with the fabrication process, including, ideally, a continuous monitoring of the symmetry as the fiber is drawn from the melt. Sometimes the preform is spun rapidly

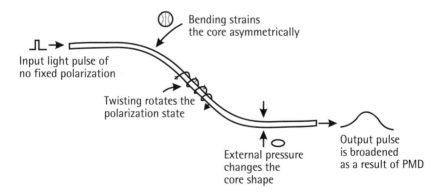

Bending strains
the core asymmetrically

Input light pulse of
no fixed polarization

Twisting rotates the
polarization state

External pressure
changes the
core shape

Output pulse
is broadened
as a result of PMD

Figure 7.19 Some practical causes of PMD.

as it passes into the oven in order to try to average out the asymmetries as much as possible.

Research on PMD is continuing, but it is more of a problem when trying to increase the bandwidth of existing links, where it was not originally the limiting factor, than for the newly installed ones, which have benefited from the improvement in fiber fabrication, following from the greater recent awareness of the importance of PMD.

7.7   Nonlinear optics and solitons

Finally, we come to the front edge of the technology and look into the next generation of systems. To get even a glimpse of what is happening in this domain, it is necessary to look at some ideas that have been covered only peripherally in the previous chapters. These ideas relate to what is called nonlinear optics.

Some readers may have been wondering why much more optical power than only a few milliwatts has not been launched into fibers in the systems we have described, in order to overcome the problem of fiber attenuation over long distances. The reason lies in the fact that the light wave actually disturbs the fiber material as it passes along the fiber core. One manifestation of this disturbance is the fact that the light travels more slowly in any material medium than it does in a vacuum—the material clearly is offering some resistance to the light wave. At the atomic level, this is because the electric field of the wave is trying to get the atomic electrons to move, and they are bound to their atoms by the attraction of the nucleus. However, provided that the light wave is not too powerful, the electrons just vibrate quite gently within the atoms, and are happy to let the wave pass with only token impediment, the magnitude of which is manifested by the change in the speed of light in the medium; that is, by the value of its refractive index (Figure 7.20a). At this amplitude, the wave in a typical monomode optical fiber is at the level of the few milliwatts with which we have become quite familiar.

If the power level rises markedly above these values, however, the electrons become agitated, and their movements in the atoms become quite strained. They are not now just vibrating gently, but quite violently, and not at all smoothly. The result is that they themselves start radiating optical power at wavelengths determined by their own local conditions, and these wavelengths may not be the same as the driving wave that is causing their

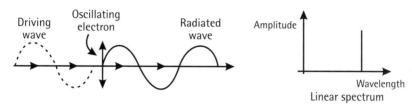

(a) Radiation from a smoothly-oscillating electron

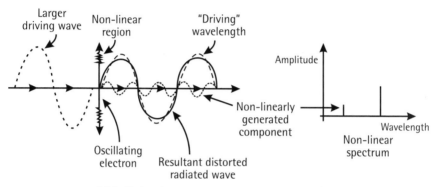

(b) Radiation from a non-linearly oscillating electron

Figure 7.20 Nonlinear optics.

disturbance (Figure 7.20b). A simple analogy is that of an electric spark—a highly nonlinear event—that radiates power over a broad range of wavelengths, from X rays to radio bands. Lightning, for example, can both be seen (optical), and heard on a radio receiver (radio waves).

The basic mathematics describing the nonlinear generation of wavelengths other than that of the driving wave are presented in Appendix E.

The result, then, is that we have, emerging from the fiber, light at wavelengths different from the one which we injected, and which also contains some of the signal modulation. So the increase in launched power has not increased the power level at our chosen wavelengths; instead, the extra power has been transferred to other wavelengths. We have a nonlinear optical phenomenon (Figure 7.20b). Another way of understanding this is to realize that the oscillation of the radiating electron is no longer a pure sinusoid, so that other wavelengths must now be present. We know this from the discussion of Fourier synthesis in Section 1.5 (Figure 7.20b).

There are several disadvantages of all of this. All our attenuators, optical amplifiers, dispersion compensators, couplers, filters, photodetectors, and so on, have been optimized for our chosen wavelength, not for any other. System performance therefore is degraded rather than improved. And there are obvious implications for WDM systems, where each channel might generate wavelengths corresponding to those of some of the other channels, thus causing information on one channel to be transferred to another, and hence leading to troublesome cross-talk between channels. It is for these reasons, then, that there are limits on the power levels that we can inject into the fiber.

However, just as there is no technical advantage without some disadvantage, so the reverse is also true. The disadvantage of optical nonlinear effects in fibers can be turned to advantage and, indeed, might lead, over the next twenty years or so, to a new generation of high-speed communications systems. What the present-day research is trying to do in this regard is to use nonlinear optics to overcome all the dispersion in the fiber, over any distance, so that global links can become truly dispersion-free.

To understand what is happening here, consider a single pulse such as is shown in Figure 7.21a. We know that the pulse must come from a laser source that has a certain spectral width, and which thus contains a range of optical wavelengths. We also know that any material dispersion in the fiber will cause these wavelengths to travel at different speeds, thus broadening the pulse (Figure 7.21a). Let us assume that the shorter wavelengths travel faster than the longer wavelengths; this is called negative dispersion, as we know. We know also that the dispersion can be either positive or negative (or, exceptionally, zero) according to the local shape of the refractive index variation with wavelength.

If, now, the pulse has quite a large amplitude (i.e., it represents a large value of power at its peak value), then nonlinear effects set in, and other optical wavelengths are generated. Some of the pulse's power is transferred to these new wavelengths (Figure 7.21b). A detailed study of the generation mechanism (well beyond the present level of description) shows that wavelengths longer than the center frequency of the spectrum are produced when the power level is rising (at the front of the pulse), and shorter ones when the level is falling (at the back of the pulse). This means that the negative dispersion will now have the effect of compressing the pulse (Figure 7.21b). It is not difficult to imagine that the two effects—material dispersion broadening and nonlinear dispersion compression—can be made to cancel each other out, and lead to a pulse that does not change in shape as it travels over very large distances (Figure 7.21c). This is indeed the case. Carefully arranged conditions can lead to just such a pulse, which is

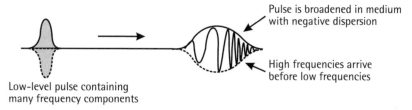

Pulse is broadened in medium with negative dispersion

High frequencies arrive before low frequencies

Low-level pulse containing many frequency components

(a) Pulse broadening as a result of negative dispersion

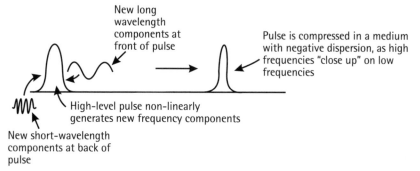

New long wavelength components at front of pulse

Pulse is compressed in a medium with negative dispersion, as high frequencies "close up" on low frequencies

High-level pulse non-linearly generates new frequency components

New short-wavelength components at back of pulse

(b) Pulse compression as a result of non-linearity in a negative-dispersion medium

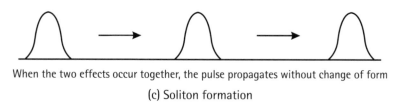

When the two effects occur together, the pulse propagates without change of form

(c) Soliton formation

Figure 7.21 Soliton propagation.

called a soliton. The soliton can, under certain conditions, even be impervious to PMD.

Solitons are not confined to optical systems; they occur in all types of wave motion, wherever there is nonlinearity in a dispersive medium. They were first noticed (without being understood) in Scotland in 1834, when one was observed to be traveling in the Union Canal, which links Glasgow and Edinburgh.

Solitons thus provide the potential for terabit technology, even without using DWDM. With DWDM, we begin to catch sight of petabit (10^{15} bits per second; i.e., 1,000 Tb.s^{-1}) technology.

Solitons overcome the dispersion problem, while the optical-fiber amplifier overcomes the attenuation problem. With such technology in the future, the capacity of high-speed global telecommunications can transform the ways in which we communicate and interact across continents.

7.8 Summary

In this chapter we have looked at some of the more detailed applications of the principles described earlier in the book. These applications include several of the more advanced components that our system design studies have shown to be needed for improved performance: couplers, switches, amplifiers, and compensators.

We have also glimpsed the future of the technology by looking at how present-day research, especially in optical amplifiers and nonlinear optics, is aimed at removing all limitations of attenuation and dispersion from the next generation of systems, to allow another transformation in the structures of global human communication.

8

The Future

This book has been concerned with understanding optical-fiber telecommunications. A wider understanding of the principles that underlie the operation of this technology is necessary to ensure that its technical development proceeds at an appropriate rate, and that it provides optimum benefit for humankind. A better understanding of these principles must lead to a correspondingly better understanding of its future capabilities and of its dangers.

Let us expand a little on these points. First, how do we expect the technology to develop? Clearly, the trend is toward channels of greater and greater bandwidth. This extra bandwidth will allow more and more information to flow and to become available to more and more people around the globe. This will necessarily change the way we do business and the way we use our leisure time. But the information age, ushered in by inexpensive computers together with satellite, mobile radio, and optical communications, has only just begun. The real advantages of universal access to the world's knowledge and activity bases are only just becoming apparent. Adjustments will need to be made. Knowledge and expertise once the jealously guarded property of governments, institutions, companies, and professions will become much more widely available. This presents dangers as well as benefits. Again, a good understanding of the technology will help us thread our way through the difficulties.

Second, what do we need in order to develop the technology? Probably the next most important advance will be fiber to the home. Fiber channels directly into each private residence will provide the bandwidth not only for access to the world's information flows from the comfort of an armchair, but also will allow the home dweller to feed information into these flows and expect a response. This will produce a sea change in the complexion of our society. We will become much more mutually interactive, with each other and with our institutions.

Advances such as this will require an increase in bandwidth of at least two, and possibly three, orders of magnitude (i.e., × 1,000) above what is available at present. Only optical fibers are capable of providing this in the foreseeable future.

Throughout the chapters of this book we have caught glimpses of what needs to be done to achieve such technology over, say, the next 10 years. Nonlinear optics allied to DWDM appears capable of providing up to 10^{15} bits per second (i.e., one petabit per second) of channel capacity in a single fiber. To engineer such systems it is necessary to develop a whole range of associated devices: the fibers themselves, fast tunable lasers, fast detectors, amplifiers, couplers and switches, storage and processing technologies, advanced networking, and so on. Research aimed at providing all of these is in active progress but, importantly, it is presently manpower-limited. There simply aren't enough people with the required skills at the time of writing (2000). Hence it has become a problem for society as a whole, and has entered the political arena. If the collective desire and will are there, resources can and should be suitably diverted. Clearly, however, the possibilities offered by the technology must be both understood and valued by society at large if real progress is to be made.

What is likely to be the technological path toward these objectives? In order to develop the fast devices and systems up to petabit (or even up to terabit; i.e., 10^{12} bits per second) rates, the emphasis has to be on the discovery and development of new optical and optoelectronic materials. Already, recent research has thrown up new optical polymers, new semiconductors, and the layered materials known as multiple-quantum-well (MQW) materials. (These latter are especially interesting, because they provide the means by which materials scientists can actually tailor the material's optical properties, by laying down layers of material at the molecular level in a carefully controlled way). Of course, such materials have applications in other important areas, not just in telecommunications: examples of these are medicine, transport, entertainment, display, advertising, and retailing. All of this will add momentum to change in lifestyles.

In the longer term, careful thought must be given to the manipulation and use of the vastly increased access to, and flow of, information. We must learn anew how to generate information, how to interpret it, how to switch it, store it, protect and restrict it, filter it, process it, and direct it to where it is most needed. These requirements will necessitate the development of a global communications network of a kind that is fundamentally different from what is presently in existence. The new kind of network will probably make use of processing that itself takes place at the optical level. This means using photons rather than electrons as the processing components in computers, and it has important advantages in that photons, unlike electrons, possess no electric charge, so do not disturb each other when in close juxtaposition. This allows for easy parallel processing, whereby many interrelated computational tasks can be executed simultaneously, hence massively increasing computing speed and power. The human brain is an example of a very powerful parallel processor. The enormous capacity of the human brain, even human intelligence itself, appears to be the result of the highly complex way in which its basic switches, its neurons, are interconnected, rather than of the speed at which they operate (which is quite slow: about 0.5 ms). Man-made networks based on such ideas are often called "neural nets." The optimum interconnection of nodes in a global telecommunications network almost certainly will owe much to the brain's methods of operating, and there exists the possibility that such a network will eventually develop an "intelligence" of its own, whatever that may truly mean.

The technological development of optical-fiber telecommunications thus carries with it great responsibilities for all involved. It is the task of the scientist to discover and elucidate the relevant principles; it is the task of the engineer to implement those principles in the design of devices and systems. It is the task of members of society as a whole to decide the extent to which the technology is needed, how quickly it can develop, what material benefits it can provide, and how quickly society can adapt to the changes it brings in its train. Society as a whole must, therefore, play an important part in the process. The more widely understood the technology, the more effective will be its contribution.

There is much that remains to be done. The future looks exciting indeed.

Appendix A:
Fourier Synthesis

Consider a periodic variation in time, t, such as is shown in Figure A.1 (the waveform shown is known as a square wave). Let us represent this function by $f(t)$, and suppose that the period of time after which it repeats itself is T. The sinusoid most closely matching this variation will have frequency $1/T$, equal to $\omega_0/2\pi$, say (Figure A.1). The function can be represented by the sum of an infinite number of synthesizing harmonics of ω_0; that is:

$$f(t) = a_0/2 + \sum_{n=1}^{\infty} \left(a_n \cos n\omega_0 t + b_n \sin n\omega_0 t \right)$$

where $a_0/2$ is a DC term; that is, a component that does not vary with time. These harmonics (including the DC term) are the Fourier components, and the process of adding them to produce $f(t)$ is known as Fourier synthesis.

The a_n, b_n can be determined by multiplying both sides of this equation by the corresponding sinusoid and then averaging over the period T. For example, to find the amplitude of one of the components, a_m, say:

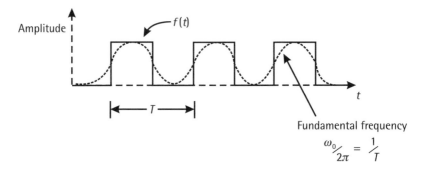

Figure A.1 Definition of $f(t)$ and its fundamental Fourier component.

$$f(t)\cos m\omega_0 t = a_0/2.\cos m\omega_0 t$$
$$+\sum_{n=1}^{\infty}\left(a_n\cos n\omega_0 t\cos m\omega_0 t + b_n\sin n\omega_0 t\cos m\omega_0 t\right)$$

On the right-hand side of this equation, using well-known trigonometrical identities, we have three expressions:

1. $a_0/2.\cos m\omega_0 t$

2. $\sum_{n=1}^{\infty}\left(a_n/2.\cos(n+m)\omega_0 t + a_n/2.\cos(n-m)\omega_0 t\right)$

3. $\sum_{n=1}^{\infty}\left(b_n/2.\sin(n+m)\omega_0 t + b_n/2.\sin(n-m)\omega_0 t\right)$

The average value of any pure sinusoid, over an integral number of periods, is zero. Consequently, all the synthesizing sinusoids, being harmonics of $1/T$, will average to zero over the period T. Hence, the only term that survives the averaging process is the cosine term, in expression 2, above, for which $m = n$. The second term in summation then becomes $a_m/2$.
Performing the averaging process on both sides:

$$1/T\int_0^T f(t)\cos m\omega_0 t\, dt = a_m/2$$

(Remember that an integration is simply a sum of infinitesimals, and hence dividing by the interval, T, over which the sum occurs will provide an average value of $f(t)$ over this interval).

Thus, in general, we have:

$$a_n = 2/T \int_0^T f(t)\cos n\omega_0 t \, dt$$

$$b_n = 2/T \int_0^T f(t)\sin n\omega_0 t \, dt$$

Now $f(t)$ can also be expressed in the form:

$$f(t) = a_0/2 + \sum_{n=1}^{\infty} A_n \sin(n\omega_0 t + \phi_n)$$

where

$$A_n^2 = a_n^2 + b_n^2$$

and

$$\tan\phi_n = b_n/a_n$$

This allows a Fourier spectrum to be represented as shown in Figure A.2, with A_n and ϕ_n plotted as a function of frequency $n\omega_0$.

Modulation bandwidth

Suppose that we have a carrier wave:

$$\sin \omega_c t$$

and we wish to modulate its amplitude by a signal wave:

$$a_s \sin \omega_s t$$

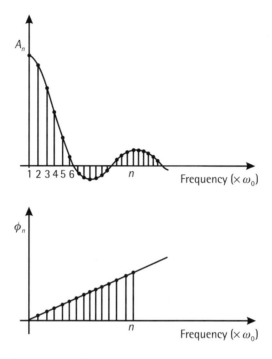

Figure A.2 Fourier spectrum for a square wave.

This can be done by allowing the amplitude of the carrier to vary in sympathy with the signal, between certain limits (see Figure A.3a). The modulated wave can be expressed in the form:

$$A_m = \left(1 + a_s \sin \omega_s t\right) \sin \omega_c t$$

The amplitude of the carrier now varies sinusoidally between a maximum value of $(1 + a_s)$ and a minimum value of $(1 - a_s)$, at a frequency ω_s.

Using a trigonometrical identity we can write the modulated wave in the form:

$$A_m = \sin \omega_c t - a_s / 2 . \cos\left(\omega_c + \omega_s\right)t + a_s / 2 . \cos\left(\omega_c - \omega_s\right)t$$

Hence, sidebands (i.e., other, flanking, frequency components) have been created at frequencies $(\omega_c + \omega_s)$ and $(\omega_c - \omega_s)$. The spectrum of the modulated signal thus takes the form shown in Figure A.3b.

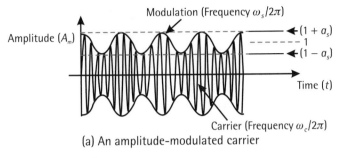

(a) An amplitude-modulated carrier

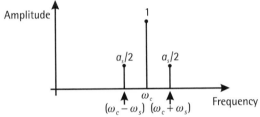

(b) Amplitude spectrum of sinusoidally-modulated carrier

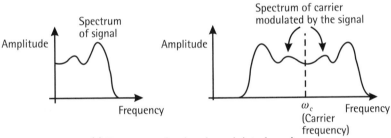

(c) Spectrum of a signal-modulated carrier

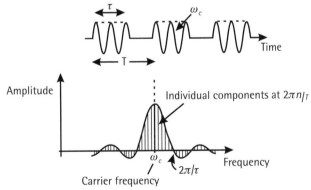

(d) Amplitude spectrum of a square-wave-modulated carrier

Figure A.3 Spectra of modulated carriers.

In general, any given signal will be represented by a waveform with a spread of Fourier components. When modulated on to a carrier, each of these components will produce two sidebands, as above, so that the width of the spectrum of the modulated carrier will be twice that of the signal itself, symmetrically arranged around the carrier (see Figure A.3c).

For example, a carrier that is being switched on and off is equivalent to amplitude modulation of the carrier by a square wave, such as the one shown in Figure A.1. Its spectrum is, therefore, that shown in Figure A.3d.

Appendix B:
The Sampling Theorem

The sampling theorem may be stated in the following form: If an analog waveform has a bandwidth B Hz, then it can be specified completely by sampling its value at a rate of $2B$ samples per second. This means that if the waveform's value is taken at specific points in time at that rate, enough information will be available to reproduce the waveform exactly.

The proof is as follows.

Suppose that the waveform $v(t)$ has duration T. Suppose now that we consider the waveform to be repeated at intervals of T (Figure B.1a). Let the bandwidth of the waveform be B Hz. From Fourier theory the repeated waveform, periodic in T, can be represented as the sum of harmonics of the fundamental frequency, $2\pi/T$. The repeated waveform can, therefore, be represented by:

$$v'(t) = \sum_n \left(a_n \cos 2\pi nt / T + b_n \sin 2\pi nt / T \right) \qquad \text{(B.1)}$$

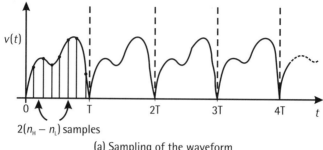

$2(n_H - n_L)$ samples

(a) Sampling of the waveform

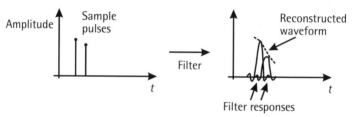

(b) Reconstruction of the waveform from the sample pulses

Figure B.1 Waveform sampling and reconstruction.

where the a_n, b_n are constants that specify the amplitude and phase of the nth harmonic of the fundamental frequency.

Suppose that the lowest frequency present in the waveform is f_L and the highest f_H. Then the lowest (n_L) and highest (n_H) values of n that are necessary are given by:

$$n_L/T = f_L; \quad n_L = f_L T$$

$$n_H/T = f_H; \quad n_H = f_H T \tag{B.2}$$

where n_L and n_H are, of course, positive integers. Hence the total number of harmonics necessary to define $v'(t)$ in (B.1) will be $(n_H - n_L)$.

Since there are two unknowns, a_n, b_n, for each value of n, it follows that $2(n_H - n_L)$ independent samples of $v'(t)$ will be sufficient to set up $2(n_H - n_L)$ linear equations for the determination of a_n and b_n for each n.

But, from (B.2):

$$2(n_H - n_L) = 2(f_H - f_L) T \tag{B.3}$$

If the $2(n_H - n_L)$ samples are taken in the time T they will be sufficient to define the waveform completely, since it merely repeats after the interval T. To take $2(n_H - n_L)$ samples in time T implies, from (B.3), a sampling rate:

$$f_s = 2(n_H - n_L)/T = 2(f_H - f_L) = 2B$$

which proves the theorem.

Note that if the waveform's spectrum extends down to DC, then the sampling rate will be just $2f_H$; that is, twice the highest frequency present in the signal waveform.

Maximum accuracy will be achieved for sampling at regular intervals within T, for this will provide the best "conditioning" for the matrix (a_n, b_n). Indeed it is possible for the accuracy of determination of some of the a_n, b_n to collapse to zero under some conditions. For example, if one of the frequency components, say $2\pi n/T$, were sampled at intervals equal to the period T/n, then only one piece, rather than two pieces, of information is available for it and hence a_n and b_n cannot be separately determined. Provided, however, the sampling interval is equal to $T/2n_H$—that is, sampling is done uniformly at a frequency equal to twice that of the highest frequency present—then the above cannot occur for any lower frequency, and the matrix is well-conditioned.

The waveform can be reconstructed, in practice, from the sample pulses, by passing the resulting pulse train through a filter of the same bandwidth, B Hz. In this case the sum of the filter's responses to the pulses reproduces the waveform (Figure B.1b).

Clearly, a knowledge of the a_n, b_n also allows the waveform to be computed directly.

Evidently, f_s represents a minimum rate and the accuracy can be improved by sampling at a rate greater than f_s, a procedure known as oversampling.

Appendix C:
Shannon's Theorem

Shannon's theorem provides an exact expression for the rate at which information in bits per second can be passed along a given communications channel when the signal level, the noise level, and the bandwidth are known. It thus allows the relationships between all of these quantities to be defined exactly. Shannon constructed a rigorous proof that employed multidimensional spaces [1], but a much simpler argument gives the same result, and keeps a firm grasp on the basic ideas. This latter argument is as follows.

Suppose that we have a sample of a signal of magnitude S_v volts and a signal bandwidth of B Hz. Suppose also that the noise level on the signal has an average level of N_v volts. Under these conditions it will not be possible to specify the signal to an accuracy of better than N_v volts so that, effectively, any S_v can only be specified to one of S_v/N_v distinct voltage levels (Figure C.1). To these must be added the 0 volts value, so that the total number of possible, specifiable levels for S_v becomes:

$$m = (1 + S_v / N_v) \qquad \text{(C.1)}$$

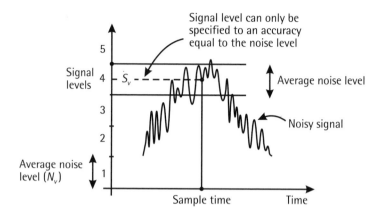

Figure C.1   Effect of noise level on accuracy of signal level.

Now, S_v/N_v relates to what we usually refer to as the signal-to-noise ratio (SNR). It is normally expressed as a ratio of electrical powers rather than voltages. Since the power ratio is simply the square of the voltage ratio (in a fixed resistor), we have:

$$SNR = S'/N = S_v^2/N_v^2$$

where S' is now the sampled signal power and N the average noise power. If we now (reasonably) assume that

$$S' \gg N$$

then it follows from (C.1) that:

$$m = (1 + S_v/N_v) \cong (1 + S'/N)^{1/2}$$

From Appendix B, the sampling theorem, we learn that a signal of bandwidth B Hz can be accurately reproduced provided that it is sampled at a rate of at least $2B$ samples per second. If two successive samples have numbers of distinct levels given by m_1 and m_2, then the total number of distinct combinations of levels that the two samples can represent is $m_1 \times m_2$.

Correspondingly, for n samples with numbers of distinct levels:

$$m_1, m_2, m_3 \ldots \ldots m_n$$

The total number of combinations will be:

$$M = \prod_{s=1}^{n} m_s$$

that is, the continued product of the m_s.

Now, we know that we need B samples per second and hence, in a time T seconds, we will need $2BT$ samples. Each sample contains, on average:

$$m = (1 + S/N)^{1/2}$$

levels, where S is now the average signal power. S/N, the ratio of the two average powers, is now the SNR as commonly expressed in electronics.

Hence, in time T, the total number of combinations of levels that can be represented is the continued product:

$$M_T = (1 + S/N)^{1/2.2BT} = (1 + S/N)^{BT}$$

This number can be represented in digital code by a number of bits given by:

$$b = \log_2 M_T = BT \log_2 (1 + S/N)$$

Hence the number of bits that can be transmitted per second—the channel capacity—is given by:

$$C = b/T = B \log_2 (1 + S/N) \qquad (C.2)$$

This is Shannon's theorem.

It shows, among other things, that for a given channel capacity, bandwidth can be exchanged for SNR, and vice versa, according to (C.2). For example, in order to transmit a given number, C, of bits per second with a lower value of S/N (i.e., a lower SNR), it is necessary to increase the bandwidth, B, and vice versa.

It should be remembered, however, that noise is usually dependent upon bandwidth. If the noise power is simply proportional to bandwidth (as is the case for shot noise and thermal noise, for example) then (C.2) can usefully be written:

$$C = B \log_2 (1 + S/KB)$$

where K is a known constant for the channel. This is an equation that can readily be solved for C, B, or S for any given set of circumstances.

Reference

1.   Shannon, C.E., "Communication in the Presence of Noise," *Proc. IRE*, Vol. 37, January 1949, pp. 10-21.

Appendix D:
Basic Theory of Laser Action

Let us consider a three-level laser. The energy-level diagram for the laser medium is shown in Figure D.1.

Pump light with frequency ν_{13} will raise the atoms from level 1 to level 3. The atoms then relax quickly to level 2, which is a metastable level. An inverted population thus builds up between levels 2 and 1. Any photon of

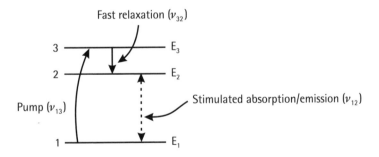

Figure D.1 Laser energy-level diagram.

frequency ν_{12} impinging on this system will now cause the stimulated emission of another photon. Our first task is to find an expression for the factor by which the light intensity exiting such a system increases over the input intensity.

Let us take a volume of material with an existing inverted population as a result of pumping. Let us assume that, at equilibrium, there are N_1 atoms per unit volume at energy level 1 and N_2 at energy level 2.

Let us now assume that light of frequency ν_{12} enters this medium and, at the point of entry, there are p photons per unit volume with frequency ν_{12}. These photons will do two things: some will be absorbed in raising atoms from energy level 1 to energy level 2; others will cause stimulated emission from energy level 2 to energy level 1, thus creating further photons of frequency ν_{12}. The total number of such photons created per unit volume per second will be the difference between the rates of these two processes. Each rate is proportional both to the density of photons and to the densities of the respective states, hence:

$$dp_a/dt = -\sigma p N_1 \qquad \text{(for absorption)} \qquad \text{(D.1a)}$$

and

$$dp_s/dt = +\sigma p N_2 \qquad \text{(for stimulated emission)} \qquad \text{(D.1b)}$$

where σ is a constant for the energy transition $1 \Leftrightarrow 2$.

Hence the rate at which photons increase in number, per unit volume, is given by the sum of (D.1a) and (D.1b):

$$dp/dt = \sigma (N_2 - N_1) p \qquad \text{(D.2)}$$

Now, suppose that the input light is passing in direction Ox through the material. Its velocity is

$$dx/dt = c$$

where c is the velocity of light in the material at this frequency, ν_{12}.

Hence from (D.2):

$$dp/dt = dx/dt . dp/dx = c \, dp/dx = \sigma (N_2 - N_1) p$$

or

$$dp/p = \sigma/c. \, (N_2 - N_1) \, dx$$

Integrating this equation gives us the way in which p increases with x:

$$p = p_0 \exp \{\sigma/c. \, (N_2 - N_1) \, x\} \qquad\qquad (D.3)$$

where p_0 is the photon density at the input point (i.e., at $x = 0$).

Note that this is only true if N_1 and N_2 are independent of x; in other words, if the inverted population is not seriously disturbed by the passage of the light. This, in turn, is only true if the input light has a much smaller effect than the pump light. We are thus dealing with "small-signal" conditions.

Now, p and p_0 are proportional to the intensities (i.e., powers per unit area) of the light at their respective positions (i.e., at x and at $x = 0$) at frequency v_{12}. In fact:

$$p = I \,/\, hv_{12} \, c$$

where h is Planck's constant.

Hence (D.3) can be written:

$$I = I_0 \exp \{\sigma/c. \, (N_2 - N_1) \, x\}$$

So the intensity of the light increases exponentially with x as the light progresses through the medium, with exponential gain factor:

$$g = \sigma/c. \, (N_2 - N_1) \qquad\qquad (D.4)$$

The actual value of σ, from the full theory, is given by:

$$\sigma = B_{12} \, hv_{12}$$

where B_{12} is one of the so-called Einstein coefficients.

Suppose this material is now used as a laser medium; for example, it is placed in a cylindrical container of length L, with parallel mirrors at each end, and is pumped to give an inverted population (Figure D.2). We can

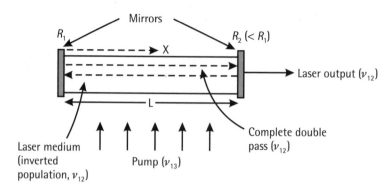

Figure D.2 Definition of laser parameters.

now use (D.4) to derive the condition necessary for lasing to occur in this arrangement.

The medium and the container will give rise to scatter and absorption losses (other than those due to the $1 \rightarrow 2$ transition) and these will lead to an attenuation of intensity with distance between the mirrors, of the usual Beer's law form:

$$I = I_0 \exp(-\alpha x)$$

There will also be losses owing to the reflectivity of the mirrors, which will, in neither case, be 100%. Suppose that the reflectivities are R_1 and R_2, respectively. Then the intensity of light after one complete go-and-return pass (a distance of 2L) between the mirrors will be given by:

$$I/I_0 = R_1 R_2 \exp\{2(g - \alpha)L\} \tag{D.4}$$

The condition for laser action to occur is that:

$$I/I_0 > 1 \tag{D.5}$$

for then more photons are produced than are lost, per pass between the mirrors. Thus the condition for lasing, from (D.4) and (D.5), is:

$$\exp\{2(g - \alpha)L\} > 1/R_1 R_2$$

Appendix E:
Nonlinear Optics: Generation of Radiated Components at Other Frequencies and Wavelengths

When a driving electromagnetic wave strikes an electron in an atom or molecule, it sets it into oscillation as a result of the force exerted by the electric field on the electric charge. The oscillating electron then generates a radiated wave. Any accelerated electron will radiate an electromagnetic wave.

Let us first consider the linear regime, where the displacement of the electron from its position of equilibrium is proportional to the force applied to it by the electric field of the driving wave (Figure E.1a).

If the displacement is given by x and the electric field by E, we have:

$$x = C\ E \qquad (E.1)$$

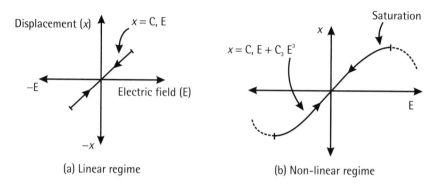

Figure E.1   Non-linear electron displacement.

where C_1 is an atomic strain constant. Suppose that the electromagnetic wave is moving in a direction Oz, has a frequency f, and a wavelength λ. Then we can write:

$$E = E_0 \sin (\omega t - kz)$$

where t is the time, $\omega = 2\pi f$, $k = 2\pi/\lambda$.
  It follows, from (E.1), that

$$x = C_1 E_0 \sin(\omega t - kz)$$

The radiated wave amplitude will be proportional to the electron's acceleration; that is, proportional to:

$$d^2x/dt^2 = -\omega^2 C_1 E_0 \sin(\omega t - kz)$$

Hence the wave radiated by the electron will be at the same frequency, f, as the driving wave ($\omega = 2\pi f$).
  Suppose now, however, that the driving wave is of much greater magnitude than for the above. It is so great, in fact, that the electron displacement is no longer linear with the applied force but begins to saturate. It has entered the nonlinear regime (see Figure E.1b). (Eventually, if the driving wave continues to increase, the electron will break free from the atom altogether and we shall have ionization).

Let us express the nonlinearity of the new relationship between displacement and electric field by adding a "cubic" term (see Figure E.1b); that is,

$$x = C_1 E + C_3 E^3$$

With, again:

$$E = E_0 \sin (\omega t - kz)$$

We have that the displacement, x, is now given by:

$$x = C_1 E_0 \sin (\omega t - kz) + C_3 E_0^3 \sin^3 (\omega t - kz) \qquad \text{(E.2)}$$

Now, a well-known trigonometrical identity tells us that:

$$\sin^3 \theta = 3/4.\sin\theta - 1/4.\sin 3\theta$$

So we have, using this in the second term of (E.2):

$$x = (C_1 E_0 + 3/4.C_3 E_0^3) \sin (\omega t\text{-}kz) - 1/4. C_3 E_0^3 \sin 3 (\omega t\text{-}kz)$$

Hence the radiated wave amplitude is now proportional to:

$$d^2 x/dt^2 = -\omega^2 (C_1 E_0 + 3/4.C_3 E_0^3) \sin(\omega t - kz) + 9/4.\omega^2 C_3 E_0^3 \sin 3 (\omega t - kz)$$

Thus, on the right-hand side of this equation, we again have a term (the first) that has the same frequency of the driving wave, but we also have the second term, which has three times the frequency (and thus one-third of the wavelength) of the driving wave. The nonlinearity has generated a "third harmonic."

It is by means such as this that nonlinearities generate alternative frequencies. Many of the mechanisms by which nonlinearities generate other frequencies and wavelengths are much more complex than this, and depend upon the independent movements of the electrons within the atom before the driving wave arrives. Such mechanisms can generate lower frequencies than the driving wave as well as the higher ones.

About the Author

Alan Rogers is a professor of electronics at King's College, London. He has published more than 190 papers in learned journals and at international conferences, and has initiated 11 patents. Professor Rogers obtained a Double First in natural sciences from Cambridge University and a Ph.D. in space physics from University College London. He has had many years of teaching and research experience, the latter ranging over radio astronomy, space physics, radio and microwave communications, optical communications, and optical measurement sensing.

Professor Rogers is a Fellow of the Institute of Physics, a Fellow of the Institution of Electrical Engineers, and a Senior Member of the Institute of Electronic and Electrical Engineers.

Index

Optical Fiber Communication Systems, Leonid Kazovsky,
  Sergio Benedetto, and Alan Willner

*Optical Fiber Sensors, Volume Four: Applications, Analysis, and
  Future Trends*, John Dakin and Brian Culshaw, editors

Optical Fiber Sensors, Volume Three: Components and Subsystems,
  John Dakin and Brian Culshaw, editors

Optical Measurement Techniques and Applications, Pramod Rastogi

*Optoelectronic Techniques for Microwave and Millimeter-Wave
  Engineering*, William M. Robertson

Reliability and Degradation of III-V Optical Devices, Osamu Ueda

Smart Structures and Materials, Brian Culshaw

*Surveillance and Reconnaissance Imaging Systems: Modeling
  and Performance Prediction*, Jon C. Leachtenauer and
  Ronald G. Driggers

Tunable Laser Diodes, Markus-Christian Amann and Jens Buus

Understanding Optical Fiber Communications, Alan Rogers

Wavelength Division Multiple Access Optical Networks,
  Andrea Borella, Giovanni Cancellieri, and Franco Chiaraluce

For further information on these and other Artech House titles,
including previously considered out-of-print books now available
through our In-Print-Forever® (IPF®) program, contact:

Artech House	Artech House
685 Canton Street	46 Gillingham Street
Norwood, MA 02062	London SW1V 1AH UK
Phone: 781-769-9750	Phone: +44 (0)171-973-8077
Fax: 781-769-6334	Fax: +44 (0)171-630-0166
e-mail: artech@artechhouse.com	e-mail: artech-uk@artechhouse.com

Find us on the World Wide Web at:
www.artechhouse.com